Robert Gr

Chieftain
Main Battle Tank
Development And Active Service From Prototype To Mk.11

This book is dedicated to Trooper Paul (Henry) Wilks 4/7 Royal Dragoon Guards 1977–1988 – died August 2012: Quis Separabit
Robert Griffin 2013

KAGERO

This book was aided through the consultation of works or illustrations (published in books or on the internet) by the following authors, and acknowledges their research or help in this project: Geoff Armstrong, Barry Beldam, Simon Dunstan, Richard Taylor and Walt Watts. We would also like to thank Peter Brown, Merlin Robinson and Tomasz Basarabowicz for their help in making this work possible.

Robert Griffin 2013

An early prototype, showing several variations to what became the final layout, with major variations on the exhaust box, turret side armour, armoured searchlight box, early NBC pack and split cupola [S. Jacklin].

Genesis: from Medium Gun Tank No.2 to Britain's First True Main Battle Tank

As the UK was finally emerging from the austere war years, it seemed to have returned to the forefront of tank technology, having finished the war (after years of under gunned and under armoured designs) with the Centurion. The first six prototypes were rushed to the German theatre of operation (under the codename Operation Sentry) in a belated attempt to gain combat experience against the Germany armour. The desired "trial combat" simply never took place because of the German collapse. Centurion flourished in the post war period and the design was adopted as the platform for the post-war replacement for all Lend-Lease, Cruiser and Infantry tank types. Between the end of the war and 1948 production of 100 Centurion Mk.1 and 200 Centurion Mk.2 proceeded, while the 20 pounder-armed Mk.3 was being developed. The Centurion soon won its spurs in the Korean War, where it amazed even the American troops by its ability to drive into locations that had previously been thought impossible for tanks. After its combat debut it returned home and was gradually upgraded with new weapons, extra armour and extra fuel capacity. It also was sold to many friendly nations (Commonwealth, NATO and Middle Eastern, with Israel being probably the most famous one). Right up to the late 1970s and 1980s nations like Sweden, South Africa and Israel proceeded with upgrade programs to their Centurions to meet their own unique requirements. Centurion also proved to have a long life with the UK forces with the last variants serving in the first Gulf War of 1991.

During the early postwar period there was every intention of replacing the Centurion design with a more powerful tank family known as the A45 Universal Tank. This quite simply was to be Britain's first main battle tank, and was intended to be the basic AFV platform to all arms (except the infantry); with planned variants ranging from normal gun tanks to flame thrower versions, and numerous other specialist vehicles. Doubts about the feasibility of the program appeared when, amongst other things, it was found that the latest landing craft could not take certain versions of A45. Concerns about the proposed 20-pounder armament being justified versus the weight of the A45 added further to the Universal Tank's troubles. When the Centurion Mk.3 was found to be a feasible platform for the 20-pounder, one thing led to another and finally the project was cancelled in 1949, although it had laid the groundwork for what was to became the Heavy Gun Tank 120mm Conqueror FV214.

Following the rejection of the A45 concept, the British Army adopted the policy of maintaining two types of tanks (Heavy Gun tanks and Medium Gun Tanks) to face the threat posed by the Soviet types then in use. The Soviets were known to be fielding the proven T-34/85, the T-44 and a new medium tank, the T54. These were backed up by the IS-2 and IS-3 heavy tanks, which had so deeply impressed the Anglo-American officers who witnessed the Berlin Victory Parade of 1945. The Centurion and the FV214 Conqueror Heavy Gun Tank combination lasted for about ten years after the latter entered service in 1954 on the basis of the A45/FV201 design. During this time the evolution of the Centurion continued throughout the 1950s into a highly successful and widely exported design.

The salient improvements executed upon the basic Centurion design included the replacement of the 20 pounder with the 105 mm L7 gun in 1959, which made the Centurion capable of taking out target at much longer ranges, a task that had been the province of Conqueror. The upgrade programs provided for the Centurion in the early 1960s, as well as the projected in-service date for the Centurion's own replacement led to the demise of the Medium Gun tank/Heavy Gun Tank combination, with the retirement of the Conqueror in 1966. Very briefly, the Centurion was "Queen of the Battlefield" in the British Army of the Rhine's armoured regiments, while it awaited its own replacement.

Many thought that with the Centurion's inbuilt longevity there would be no need to worry about a replacement for many years when the Mk.3 appeared in 1949, and again when the L7 was first fitted in 1959. The truth was that in 1951 a replacement program for Centurion was instigated under the title of Medium Gun Tank No. 2. While this time frame may seem a little premature, it has to be born in mind that the whole A45/FV201 program was an indication that many considered the Centurion specification inadequate to meet the threats expected from the USSR. Development of a tank takes time, effort and most importantly money, with the treasury always complaining about rising costs. The other major issue was the rapid advance in technology in the immediate post war era, so that what was state of the art one day was obsolescent the next. In the modern context one has only to look at the mobile phone industry, where virtually overnight equipment becomes

The last two Chieftain prototypes were built to Mk.1 standard for the Bundeswehr. Following trials the two Chieftains were sent to the United States: one was displayedat Aberdeen Proving Ground, one at the Patton Museum [Fort Knox].

obsolete. The specification for Medium Gun Tank No. 2, victim of the rapid technological changes of the 1950s and subject to the nuclear doctrines that would characterize NATO's defense plans for Western Europe, was delayed repeatedly while many avenues of AFV technology were explored and evaluated.

The United Kingdom was still recovering from the effects of the Second World War and was in the process of establishing

Head on shot of a Mk.2 located at Bovington Camp. This shows some of the Mk.2 features very well such as the large single headlights and just visible the jerry can holder on the left of the turret as we view the tank [J. Hall].

Close up view of a prototype turret, towards the front can be seen the remains of the clamps that held the very large canvas cover that was such a feature during the early days, this was added for security purposes. Also note the original gunner's sight hood with no armoured flap to lower over the head, split cupola and the open IR searchlight door [S. Jacklin].

An excellent view from the turret looking at the drivers compartment minus his seat, in the front centre the large foot pedal is the footbrake and to its right is the accelerator. The large levers centre are the steering levers and forward of those are the hydraulic start controls (left) and the generator controls (right). As can be seen either side of the driver are five round stowage for HESH and below those are the vehicle batteries [R. Griffin].

a welfare state at the time that Medium Gun tank No. 2 was being conceived, thus it was always hoped that money could be saved. This resulted in several weird and fanciful ideas being evaluated (with great seriousness) in order to save money on the new tank. These ideas would nowadays seem amusing, but some of them; although woefully inadequate and dangerous to the crews, were serious contenders for the specification of Medium Gun Tank No. 2. These included a tank with a limited traverse main armament, which enabled the whole profile to be lowered considerably but created tactical problems in fighting such a vehicle. Other proposals include towed trailers and armoured two part vehicles on the Snow Cat principle. These all thankfully came to nothing, but one proposal that was taken very seriously was the FV4401 Prodigal, which sometimes is linked to a vehicle in the Bovington Tank Museum known as Contentious, although they are two different vehicles.

The FV4401 adopted the idea of the one man tank, which could be built cheaply and in greater numbers than conventional tanks, additionally adding the virtue of air portability. In this aspect the concept was adequate, but the big flaw with a crew of one was overwork. The crewman would have to drive, command, aim the gun and operate the radios, a set of tasks guaranteed to wear him out very rapidly. A two man crew was later envisaged to improve the concept. Prodigal was a low, compact vehicle reminiscent of the later M50 ONTOS, and it mounted two 120 mm recoilless rifle that were based on the then-current Burney recoiless designs that would enter service in the British Army in the late 1950s. Each gun was provided with a rotary seven round magazines. Normally vehicle pro-

grams like this are designated to the "paper tanks" file, but the FV4401 was actually built and possibly in several variants for testing purposes. It was shrouded in secrecy, as evident from a minute that has survived in the National Archives, relating to an open day at Chobham (which at the time was the Armoured Vehicle Development Centre for the British Army). The minute goes into detail about how far visitors would be allowed from FV4401 and the need not to attract too many visitors in the first place. One can only think that the easiest way around this would have been to not display the vehicle. The Prodigal went through several development phases before the final version described above was finalised, and useful information was obtained on trunnion stress from large calibre weapons on such a small light vehicle (this issue would be revisited by the Americans years later during M551 Sheridan development and the Prodigal data contributed to the later CVRT program in the United Kingdom). However well the project was going, wiser heads decided that it was not the way forward for the Medium Gun Tank No. 2, and eventually the orphaned FV4401 project was cancelled. The on-going mystery is what became of the prototype vehicles (at least three having been built), no record of their destruction seems to exist and they have not turned up, so maybe somewhere in a forgotten hangar lurk three FV4401. Possibly the only tangible link is the vehicle in Bovington Tank Museum called Contentious. This may or may not be part of the FV4401 program but remains an enigma all on its own. It is aimed in the same manner as the Swedish S tank but is built from assemblies from other current UK vehicles of the time. There is very little documentation for the vehicle, so again

The IR doors on the prototype, it is interesting to compare these to a service vehicle, note how these have two doors hinged in the middle as opposed to the production version with its single door, the searchlight is also armoured whereas the production version was not. Of special interest is the use of the Centurion idler wheel, this was common on the early vehicles, along with the use of the Conqueror gun crutch [S. Jacklin].

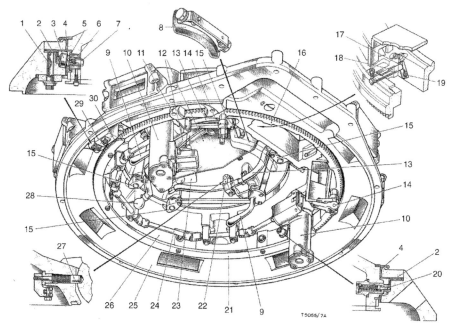

1 Lead wire	9 Cupola door locking handle	17 Catch locking pin	25 Cupola fixed ring
2 Slip ring	10 Elevating gear	18 Periscope sight locking pin	26 Cupola rotating ring
3 Roller	11 Rotating gear bracket	19 Catch spring	27 Locking pin
4 Dust cover	12 Cupola doors	20 Earthing brush	28 Elevating gear handle
5 Roller Shaft	13 Door closing handles	21 Rack guard	29 Direction sensing switch
6 Metalastic brush	14 Periscope retaining catch	22 Door locking catch	30 Nylon roller
7 Screw	15 Locking lever	23 Wiper motor	
8 Brow pad	16 Contra–rotating gear rack	24 Elevating gear drive shaft	

A workshop drawing of the inside of the original Mk.11 cupola as fitted to the Mk.1 [Crown Copyright].

1 Commander's machine gun mounting
2 Commander's periscope guard
3 Spotlight junction box
4 Periscope ocular prisms
5 Door locking handles
6 Catch release handle
7 Partly open door bolt
8 Periscope object prism retainers
9 Periscope wiper motors
10 Machine gun firing junction box
11 Periscope object prisms
12 Periscope washer jet
13 Periscope wiper blades
14 Rotating ring
15 Fixed ring

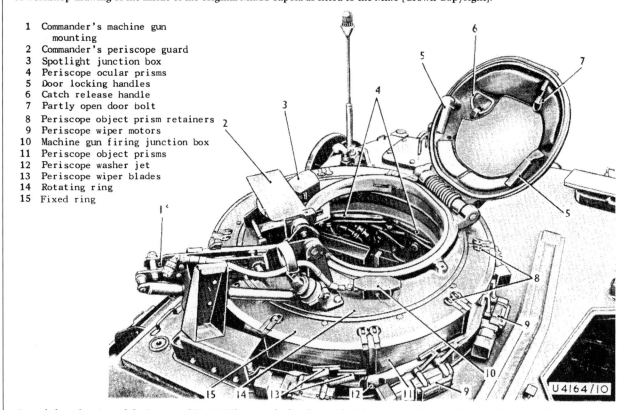

A workshop drawing of the improved No.15 Mk.1 cupola fitted to early Mk.2's, it can be seen how much better this version was compared to the Mk.11. Notice the levers located all around the top plate, these retained the top half of the split sights [Crown Copyright].

maybe one day it will emerge as a link in the Medium Gun Tank No. 2 story.

The next step in the story starts with the approval by the Army Council in 1956 to construct the Medium Gun tank No. 2 as a conventional design, the inspiration for this move was the realisation that Centurion was possibly outmatched by T-55 and it was known that a successor to T-55 was already on the drawing board. The IS-3, which had been the raison d'être for the FV214 Conqueror being in service was already being replaced by the T10. Since the Conqueror itself was only just entering service, the British feared themselves facing being outgunned again, something they had promised would never happen after the disastrous designs of WW2. To counter the anticipated successors to the T10 a guided weapon system was designed and code-named Orange William, although it was felt that long range attack on enemy armour must still be the job of the conventional gun tank.

The specification of Medium Gun Tank No. 2 was to be based around a new 105 mm gun based on a US design. Conscious of the physical effort involved in loading heavy brass cases, and based on information from loading trials on Conqueror, the new gun was to be loaded with combustible bag charges (a system well known in the navies of the world and common for large artillery pieces but never before used on a tank). This decision appeared to be correct when later in 1956 the decision was taken to replace the projected 105mm gun for Medium Gun Tank No. 2 with a new UK designed 120mm. The need for a long range killing weapon was deemed desirable to be able to engage enemy forces on the plains of Europe at maximum range, and the 120 mm was envisaged to have a longer reach than the 105 mm originally foreseen.

During this same period the United States Army was developing a futuristic medium tank of their own known as the T95. Some commonality in Anglo-American tank design proposals, such as use of T104 US 105 mm gun (which had been proposed to arm the US T95 Medium Tank) were considered. Medium Gun Tank No.2 was to have a 100-inch turret ring to allow for a large armament and a well-sloped glacis. It was intended furthermore, perhaps in the event of any future need for US production of the main armament, or in the interest of standardization, that Medium Gun Tank No.2 was to be able to fit the T95 turret and vice versa (although why anyone would want to do this remains very unclear). The T95 was a promising design, unique for its time in that its suspension units were hydraulically adjustable and surviving publicity shots of the vehicle show it in various poses.

Throughout the remainder of 1956 the Medium Gun Tank No. 2 project gathered momentum and Leyland Motors, design parent of the improved Centurion Mk.7, were tasked as main contractor role for the hull and engine of FV4201 (as the tank was now known). Because many of the concepts for the new tank were revolutionary for the time, Leyland decided to produce an interim test vehicle to incorporate some of the design features, which was hoped to have a double effect in that it would save money by testing design's paper features which had never been constructed, and it also would prove them in field trials. Leyland rapidly produced three prototype vehicles based on Centurion Mk.7 and Mk.8 components. These became quite famous in their own right and were known as 40-ton Centurions, armed with the 20-pounder, and had radically different features from their brother Centurions: five road wheels per side, a weight of only 42 tons, a reclining drivers position,

The twin hatches of the No.11 cupola showing the locking levers on the inside. Notice the angle of the loaders hatches in this shot, they run parallel to the reinforcing bar between the cupola and the hatches, later on they were offset [S. Jacklin].

A view of the early NBC system, and also of the layout of the rear turret armour. Notice the pronounced step in the armour plates, this was to become a much flusher layout in the production vehicles. The 119 REC marking relates to the Territorial Army R.E.M.E. unit who brought the vehicle from Swynerton training area to its current location at Ripon N Yorkshire [S. Jacklin].

The aperture for the .50 Browning Ranging Gun or in the case of this particular vehicle the M85 Ranging Gun. This Tank was never actually fitted with the Browning and still retains its M85 mounting, although the aperture in the turret fitted both weapons [S. Jacklin].

and a heavily sloped turret front without an external mantlet. These features were broadly mirrored in the US T95 prototypes as well, a full turret for which may have been made available to the British Army, showing that Anglo-American design trends were similar and probably informed each other at the time. The 40-ton Centurions were successfully tested alongside late-model Centurions (in 1958-59) but were never proposed as production vehicles. The 40 tonners provided much needed information before the FV4201 could go into production, and after providing this data the three tanks were abandoned; one survives at Bovington tank museum although not currently on display, one is supposed to be at the Bordon REME facility suffering the fate of being a recovery aid, and I suspect that in fact this probably has long gone by now. The third has vanished, although popular rumour has it arriving in Israel with a shipment of Centurion spares, and although Israel has never confirmed this tale it does make for an amusing image.

In 1957 the Fourth Tripartite Conference (USA, UK and Canada) established and defined the modern concept of the Main Battle Tank and set out some of the main principles for the next generation of tank design for the US Army and British Army, replacing the need for Medium and Heavy Tank classes. The main armament for any such MBT is agreed to have to be able to penetrate 120 mm of 60 degree sloped armour at 1800 meters. At this point the FV4201 design was proposed to have a V-8 Rolls Royce engine with automatic gearbox, a 120mm gun, Infra-Red night fighting equipment and weigh no more than 45 tons. The US concept remained the T95, armed with either a high performance 105 mm or a smooth bore 90 mm gun. Both tanks would still require interchangeable turrets as

a design feature, and one wonders why engine commonality was not demanded instead, especially given US expertise in the motor industry. Also one has to ask why would you swop your 120 mm turret for one fitted with a 90 mm? By now the main design features of the FV4201 were well on the way to being established; although as in all tank development, it was appreciated that changes would be required once the prototypes were up and running and even then the design team felt they had another world class tank ready and hoped for as much success as with Centurion. The Medium Gun Tank No. 2, or FV4201 would now officially be built as Britain's first Main Battle Tank.

In 1958 Vickers was brought in as the second main contractor with responsibility for turret development. The turret was to be of a new design, with minimal front cross section and dispensing with the traditional mantlet that had been in use for many years by all tank producing countries. This gave FV4201 a very compact and sharp turret front which in turn reduced the visible target area to the enemy. Other design problems that had to be overcome were efficient stowage of the ammunition; and particularly for the bagged charges, which in their original form were very susceptible to water (a commodity found in large quantities in the field). Drawing on the wet stowage concepts developed for wartime Shermans, the bagged charges were quickly provided a safe and effective stowage, although leakage in the water jacketed charge bins did occur. This same problem was to occasionally to nag the FV4201 in its early service days but in time the problem was virtually eradicated.

To further complicate the development of the FV4201, the 1958 NATO requirement for a multi-fuel MBT power plant was ominously adopted by the British Army in the aim

The rear of the turret clearly showing the two large hatches at the extreme rear. The right hand is cover under which lie the components of the powered gun kit, while the left hand hatch allows access to the turret drinking water tank and also the turret batteries. Both hatches are considerably thicker than those on the production models [S. Jacklin].

A very good shot of the prototype Chieftain that used to reside as gate guard at the vehicle research at Chertsey. Fortunately wise heads have prevailed and the vehicle is to be preserved notice how low it sits and also the slope on the rear section of the hull, this was straightened in the production versions [R. Griffin].

Showing the turret top layout of a prototype Chieftain, it can be clearly seen in this shot how the bar that gives strength to the turret roof is a straight piece but later on it had a distinctive bend and the loaders hatches were also moved [S. Jacklin].

A view of the gear box and exhaust arrangement on a prototype Chieftain. This layout for the exhaust system did not survive the prototype and was changed several times during Chieftains life [S. Jacklin].

This view of a Mk.2 shows off some of its features to good advantage, central is the bracket for carrying two jerry cans, usually one water one petrol and located underneath it one of the six round smoke grenade stowage boxes. Behind the cupola the IR detector stalk can be clearly seen in its operating position [R. Griffin].

of standardisation within NATO. Sadly this grand gesture only seemed to apply to the UK, and all the other major tank producing countries in NATO then developing their own MBTs opted for a conventional diesel (although the USA would eventually move to a gas turbine). The progress on the new Rolls RoyceV-8 for FV4201 was not going as well as expected, and a delay of two years was expected, so in the end the decision was taken to drop the V8 and go with a proposed multi-fuel engine. With the facility of hindsight, the multi-fuel decision that day doomed FV4201 to being mechanically unreliable and lost the design many future sales. The attraction of being able to change fuel type at the flick of a switch was seductive but the concept was extremely complex. It was not to be simple design change or a simple case of out with one engine, in with another. The new engine necessitated a complete rear end redesign and was the first breach of the design parameter regarding weight limit, when it was found that the new engine would push the weight to the 50 ton mark.

A design based on a WW2 German Junkers Jumo was selected by Leyland as the basis of the new engine, more for its lack of moving parts compared to similar engines than its suitability. Leyland took over the design of the new engine now designated as the L60, and it should be noted that the

end user, the Royal Armoured Corps, were dubious to say the least about the L60 (entertaining severe doubts as to the engines suitability for the new tank). Sadly these reasoned views were treated almost as heresy by the government and Leyland, and by the skilful use of its scientists the government swept all army objections under the carpet, in a shameful treatment of those whose lives might depend on the engine performing as asked.

By August 1958 the General staff had finalised what they believed they required from the new tank, namely the combined firepower of Conqueror, armour protection that would be utilised within the max permitted weight of 45 tons, top speed of 26 mph and mobility or better of Centurion. As we have seen by the time this was issued the decision of incorporating the L60 had already raised the weight. The important criteria here is the requirement for mobility to be about the same as Centurion, a decision taken while most other major tank producing counties were looking at good protection AND good mobility, another decision which would haunt the underpowered Chieftain for most of its service life.

The same year (1958) a meeting took place at Chertsey to finalise and accept the parameters decided on for FV4201, which were summed up in the wooden design mock-up on

An excellent shot of a Mk.11 Chieftain belonging to the Royal Scots Dragoon Guards, it has its green range flag up so its weapons are unloaded [Private Collection].

show at the meeting. It was to have a height of 7ft 10in, a hull length of 22ft 3in, and a ground clearance of 17in. A total of 60 rounds of 120mm ammunition were to be carried under armour and the engine was to develop 700 BHP, with a power to weight ratio of 15.5 BHP per ton. The design was accepted by the army but many modifications were listed to be incorporated into the first prototypes, which were to be numbered P1 to P6. A second order was placed in June 1959 for a further six prototypes (W1 to W6), which were built to be used eventually in the troop trials, and on top of this a further two vehicles were built for evaluation by West Germany, although these were of the later Chieftain Mk.1 standard (and both survive today in museums in the USA). Likewise the UK received two Leopard 1 pre-production

tanks and again these both survive in the UK. The prototype Chieftains (as the FV4201 would be named) drew heavily on the experience of Leyland Motors in refining the Centurion design into the Mk.7 and Mk.8, but Leyland had no experience of large diesel engines for tank applications, and in fact no British tank manufacturer involved in the 1939-45 tank programs had ever produced a high output diesel specifically for AFV applications. The power pack concept was pursued from the start in Chieftain's engine and gearbox, with the object of quick engine changes, especially after the very long engine repair jobs required in Centurion and Conqueror due to the Meteor engine's layout. The L60 was a very complex engine designed quickly for an application that even modern tank engines have yet to achieve in any comprehensive man-

Here we see a row of recently-converted Mk.11 Chieftains awaiting delivery from base workshops to their respective units. Notice the overall NATO Green paint finish: the disruptive black was added later at the unit. The pattern of camouflage applied to the Chieftain could be very uniform within a unit but following the discretion of the commanding officer, it could also show huge variation from tank to tank, or from squadron to squadron within a regiment [R. Griffin].

The Royal Scots Dragoon Guards driving down Princes Street Edinburgh, their home town, during their 1979 Tercentenary Parade. Notice for this parade for some reason all tanks have had their thermal sleeves removed from the barrels [Private collection].

ner, and there was no backup plan for an alternative power plant, a very serious flaw in the FV4201 program.

Throughout the Chieftain's service, the Royal Armoured Corps repeatedly asked for a better power plant, and the stock answer that always came back was that it could not been done, since to fit new engine would take a massive redesign of the engine compartment. This sadly was nothing more than a distortion of the truth, for eventually Chieftain *was* fitted with at least 3 different engine layouts. The Chieftain 800 & 900 test vehicles both fielded a derated Perkins diesel CV12 as used in Challenger; this fitted into the current engine bay. The Khalid used the same CV12 engine but with the inclusion of Challengers cooling system, hence the different hull shape for Khalid. An MTU engine of 1000 BHP was also fitted and tested in the 1970s and finally a gas turbine was eventually also trialled. The truth can only be that for political reasons and due to an unwillingness to assume the cost of rebuilding the entire Chieftain fleet the L60 was retained and improved, which must have eventually cost at least as much as re-engining the whole fleet would have.

Once a new AFV has been built it is subjected to many hours of hard testing which will include 24 hour road running, firing trials, cross country driving and some vehicles will be built simply to be used as firing targets to test the effectiveness of the design and armour protection. P1 and P2 were the first to receive the L60 married to the TN12 auto gear box. Because they were only ever to be used for automotive trials they never received conventional turrets, they received a large round counterweight instead to simulate the all up weight of the vehicle, and this counter weight (due its passing resemblance to a turret in Windsor Castle) was known as a "Windsor Turret". Running trials began in the early part of 1960 and during this phase it was found that to make the L60 full multi-fuel took in excess of eight hours of work, thus killing the "fill up with anything" theory held by the multi-fuel enthusiasts. The whole engine design was flawed from the start but FV4201 was stuck with the L60. During the running trials many faults began to appear with the L60, but these were overcome and it is too easy to be harsh on the L60, as some of the problems could have occurred on any new engine. The

This tank is not at BATUS although the colour scheme would suggest it, it is in fact on Salisbury Plain and belongs to the Land Warfare centre who provide "OPFOR" for units exercising on the plain, it can only be a short training period as can be evidenced by the lack of kit in the commanders basket [Plain Military].

A rear shot of a very unusual Chieftain indeed, notice the absence of the usual exhaust box and pipes. This is because automotively, this is the most powerful Chieftain ever built. Instead of the problematic L60 and TN12 power pack it was fitted with an MTU diesel linked to a Renk auto gear box. This combination made it powerful and very fast, it also had the ability to stop on the proverbial sixpence as one employee found out. Sadly although it located at a museum it is in poor condition inside, and needs a lot of work on it. The power pack is still classed as a commercial secret. [R. Griffin]

The strange contraption on the pack of the MTU Chieftain is actually a lifting frame to help lift the one piece engine deck which is not on a torsion bar unlike a similar system in Challenger. The frame is stowed when not in use so as not to hinder traverse [R. Griffin].

silencers were moved from inside the hull and placed at the rear of the hull in an armoured box, which stayed with the Chieftain design throughout its service life. The penalty for the modifications needed to make the power pack serviceable was a rise in weight, which had by now reached 50 tons. The weight increase had a knock-on effect with the suspension units having to be strengthened, which itself also added more weight. A decision was also taken to fit rubber pads to the track to enable the tank to comply with German road law, to lessen the damage the tracks had on a normal tarmac road. This again served to push the weight up until the situation arose of a power to weight ratio of only 10 BHP per ton, which was unacceptable compared to the power available to MBTs being developed by other nations. Problems with the gear boxes were causing problems as well, with only three boxes completing above 500 miles by October 1960. When a new L60 engine with 550 BHP appeared, more gear box problems started to appear in the TN12 transmission; it was noted that if the L60 had been more reliable then these issues would certainly have been discovered earlier.

While the automotive trials struggled to overcome their problems the firing trials were taking place with W3 being the first Chieftain to fire the L11 120mm in May 1961. The trials started at Kirkcudbright Ranges in Scotland and were run by FVRDE (Fighting Vehicle Research and Development Establishment), and were subsequently repeated at Lulworth

A shot of the loaders area inside the turret, right of the picture is the breech and located to the right and forward is the co-ax 7.62mm GPMG. The rear two orange levers are for opening and closing the breech while the forward one is the GPMG cocking handle. On the turret wall stowage for two APFSDS rounds can be seen with a third located vertically to the left of the picture. The large silver box is stowage for the co-ax ammunition [R. Griffin].

Hopefully this shot will dispel all the myths concerning Stillbrew armour, as described in the text it can be seen it is nothing more than steel plates backed with several layers of thick rubber [R. Griffin].

This shot is taken from directly behind the 120 mm, the breech ring is at the bottom of the shot and dominating the forward area is the recoil system which controls the gun during recoil to its maximum of 37.5 cm and then returns it to the run out position automatically opening the breech as it does. On the right upper can be seen the gunners auxiliary telescope and on the left the GPMG co-ax [R. Griffin].

ranges in Dorset under Royal Armoured Corps direction. The firing trials revealed the L11 120mm gun as an excellent design, although issues were raised in other sub systems, including the need for the replacement of the M85 .50 cal (12.7mm) Ranging Gun that had been originally fitted with the proven American M2 .50 Browning. The original M85 weapons had been specifically designed for AFV use but in practise the M85 proved to be dangerous and it could not cope with the chamber pressure generated by the current range of UK ranging ammunition.

The Ranging Gun (or RG, but never called a Ranging Machine Gun in the Royal Armoured Corps) had been introduced in the later versions of the Centurion as a simple method of finding the main armament's target range reasonably accurately at low cost. Prior to this gunnery had relied on good estimation skills by the commander, or by firing a main armament round to determine the engagement range, which was both wasteful and not tactically sound. In action the RG would be laid onto a target as part of the fire order, and once fired a burst of three special rounds would be fired, controlled by a solenoid fitted to the side of the weapon. The gunner would be looking for a strike on target, so if no strikes were observed more rounds were fired, and once strikes were observed the range was applied to the main armament and fired. Until the advent of laser range finders, the system was quick and accurate, and it also could give an indication that no other system in use can give, even today: wind direction at the target. Most modern systems will take wind direction at the firing point into account, but cannot calculate what wind direction is down range at the target.

Prototypes P3 and P4 joined in the gunnery trials which also included the use of the Infra-Red night fighting equipment in 1962, and looking back today after years of using Image Intensified Sights and Thermal Imagery for night combat, Infra-Red seems crude and very ineffective: nevertheless it was state of the art for its time. All of the prototype vehicle were constantly upgraded or modified through the trials period, and as the trials progressed and problems were gradually overcome it was felt safe to send two FV4201 Chieftain prototypes to BAOR for initial troop trials. The lucky units who would evaluate the Chieftain were both from the Royal Tank Regiment, the 1st RTR stationed at Hohne near the NATO ranges, and the 5th RTR stationed at Fallingbostel, just down the road. The vehicles selected were W1 and W3, and they were shipped to Germany in great secrecy. The crews for the trials had already attended familiarization various courses back in the UK and were ready to receive the Chieftains and to see how the new tank would perform against their beloved Centurions. First impressions were good even though the trials began in the middle of a typical cold snowy German winter. The crews found that the rubber track pads gave the Chieftain a great edge over Centurion on frozen roads, where the Centurion's steel track caused them to slip and slide. One serious issue that was raised where Centurion crews had the definite edge on the Chieftain was the total lack of crew compartment heating for the Chieftain. The Centurion had a flap that could be opened and allowed engine heat into the fighting compartment. Again the designers stated that there was no need for a heater, those same decision-makers never had to sit in an ice cold Chieftain on the North German Plain or in the middle of a BATUS winter! At one stage a few years later the

A Squadron of the 4th/7th Royal Dragoon Guards carried out trials on heated suits fitted with heated gloves and insoles for the boots. An admirable concept, but one that had not really been thought through, as all the leads were very flimsy and not suited to the rough environment of a tank, and there was no control for the heat so the wearer was either cold or red hot. The idea was discarded, although towards the end of its life heaters were eventually fitted to Chieftain.

During the cross country phase of the BAOR trials the Centurion proved better, partly due to its higher ground clearance, while Chieftain (with only 17ins clearance) was always getting bogged down. This was rectified by fitting Centurion wheels and modifying the suspension on the prototype Chieftains, which then gave a respectable clearance of 22ins. Automotively Chieftain held its' own, but due to the low power output it was only just able to so against the Centurion, which was below the Chieftain's designed level of mobility. Meanwhile the gunnery trials proceeded as planned, and as expected showed up some problems, one of which was the electro mechanical rammer that had been installed in the turret. This device was designed to ram the bagged charge as it was thought safer than the loader doing so by hand. The device malfunctioned regularly, or seemed to have a life of its own, such that it became a real and serious hazard to the loader, so the crews removed it and left it lying in the corner of the hangar. Ramming the projectile was achieved by using the charge itself pushed by the loader before closing the breech.

One advantage the Chieftain loader had over the loader on Centurion was that after firing he had no large hot brass case ejected into the turret. This was achieved by the use of combustible bag charges which were ignited by a small vent tube (a blank round) that was automatically fed into the breech ring and fired electrically. Traditional weapons using bag charges have had an interrupted thread breech with a swing open door; this obviously would be impracticable in the confines of an AFV and also too slow; so a system of metal inserts known as obturators were employed in a sliding block breech design. In its simplest form these obturators consist of two metal inserts, one in the rear of the chamber and one in the forward face of the vertical sliding breech block. To prevent damage during opening and closing of the breech the obturators were controlled by cams. Excess fumes were vented from the turret by means of a fume extractor located half along the barrel, this worked on the over pressure system and used the spent gases to evacuate the bore, and thus when the breech opened after firing no fumes entered the fighting compartment. Another innovation was the fitting of a thermal sleeve over the barrel, which was designed to help prevent "barrel bend", which occurs when the gun is firing and is subjected to varying temperatures caused by cold winds or rain.

Chieftain: Early Service and Vehicle Description

Overall the trials were seen as successful but they also showed that there was a lot more work to be done before Chieftain could be put into full production (especially on the L60 and transmission), but all these issues apart the Chieftain was accepted for service on the 1st of May 1963. This was more as an act of faith to allow the manufacturers to gear up to start

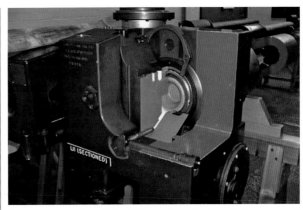

A cut away section of the 120 mm breech ring, the brass round that can be seen is the .625 vent tube which is fired electrically and the supply to that can be seen on the right running across the ring till it comes into contact with the firing needle, once fired flame travels up the whit portion of the breech block and ignites the rear of the bag charge shown here by the orange base of an APDS charge [R. Griffin].

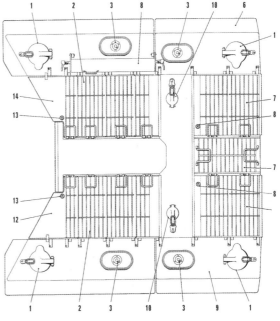

Plan of Early Chieftain Engine Decks (Mk.1, 2, 3)

1. Fuel filler cover
2. Air inlet louvers, right
3. Fuel tank breather
4. Air intake
5. Coolant filler cover
6. Right fuel tank
7. Air outlet louvers
8. Louver budget locks
9. Left fuel tank
10. Hydraulic filler cover
11. Air inlet louvers, left
12. Left front engine compartment cover
13. Cover budget lock
14. Right front engine compartment cover

User manual drawing showing the layout of the early Chieftain engine deck (typical for the Mk.1 and Mk.2), naming the parts. Four fuel fillers were provided, but during tactical replenishments only the two at the rear would be used, as the gun was always kept forward for protection [R. Griffin].

production than an admission that all issues had been solved. Despite the best efforts of the engineers only an extra 35 BHP could be obtained from the current L60 before it was used to equip the first production version of the Chieftain MBT Mk.1 of which 40 were built. These never were issued in MBT form to the front line but used only for trials sand eventually training,

From this shot of a prototype Chieftain it easy to see how it changed before coming into production, the rear top of the hull slopes, smaller road wheels and the stowage of tools on the hull side, this vehicle still carries the false front bin and large canvas cover over the turret front. These were fitted so that anyone looking at it would not have a good idea of the ballistic shape [R. Griffin].

The so-called 40 ton Centurion. Made from Centurion parts three were built to test the new ideas to be used on Chieftain, the main one being the driver's supine driving position. A similar Centurion was built using a similar turret and was known as Centurion X. Of the three the Tank Museum used to have this one is the only one left [P. Rogers].

although some of them eventually did serve as platforms for conversion to bridge layers in the 1980s. The Chieftain Mk.1 was the first Chieftain to establish the overall shape of Chieftain and it only bore a resemblance to the prototypes. Major differences existed between the prototypes and the Mk.1: these included the turret shape, the hull rear. Features like the gun travel clamp were new whereas the prototypes had used the same clamp as used by the Conqueror.

Although Chieftain would be developed and redeveloped through many marks and would eventually reach Mk.11 status in the 1980s (although some references including the Tank Museum at Bovington mistakenly list a Mk.12; this also occurs in Jane's fighting vehicles, these are incorrect. The Mk.12 was proposed and specifications issued but with the advent of Challenger 1 the project was closed). The only new factory built marks of the FV4201 Chieftain gun tank were the Mk.1, Mk.2, Mk.3 series and Mk.5 - all others were rebuilds and upgrades.

It would not be until November 1966 that the British Army received its first production vehicle Chieftain, the Mk.2, so until then the Mk.1 Chieftains were used extensively for training the future Chieftain crews. The vehicles were also modified from the initial Mk.1 and upgraded to reflect the changes seen in the frontline versions, through to the last version which was Mk.1/4.

To look at a description of the Mk.1 we first have to understand that all tank design is governed by three rules; firepower, mobility and protection. Each country will incorporate these in their designs to suit their doctrines, for example the UK prefers firepower, protection and mobility (in that order). Countries such as Germany and France have gone for firepower, mobility and protection, arguing that high speed and mobility are part of the tank's protection. There is no 100% answer to this argument and each country will do what it considers to be best in its national interest, but all three do have an impact on each other and on the final design. If for example a builder wishes to place the largest weapon feasible onto the hull, then the hull must be made large enough to carry it, which will then affect mobility and armour protection.

The hull of the Mk.1 was built of welded steel armour plate, with the floor plate shaped into a V to help increase mine protection; the hull sides were also sloped outwards in an effort to

help divert mine blast away. Although the Chieftain introduce the reclining drivers position, in many other factors it followed conventional design with the driver at the front, the fighting compartment in the centre and the rear taken up with the power pack. The upper glacis plate was cast in one large piece, centrally placed in which was the driver's hatch (which was of the "lift and swing" design). The type of hatch specified for the Chieftain's drivers position minimised the risk of the driver not being able to open his hatch when the turret was in certain positions, tragically this had not been the case on designs dating to World War 2 and drivers had burnt to death for a simple design error. Located behind at the rear part of the hatchway was the drivers day driving sight, this was used when driving closed down. It was a single wide angle periscope, giving the driver a good forward view although he still required input from the vehicle commander when reversing. This sight (Periscope AFV No36 Mk.1) could be removed and replaced with Infra-Red sight for night driving, and later on in the Chieftain's production this was changed to an Image Intensified version. Both types required a lot of training because the image produced was not a clear view as seen through normal sights. The sight was equipped with both washer and wiper systems to enable the driver to clean the sight whilst closed down.

The driver's seat was designed to allow opened up driving or driving fully closed down, which ever version was used the controls used to drive the tank were the same. To steer the driver had two steering levers instead of a conventional steering wheel. To turn left he applied the left lever and exactly the same to turn right. An accelerator pedal was located on the right of the driver's 3 foot pedals, centrally mounted was the footbrake (with enough space for both feet to be placed on it if need be). The gears were changed by the means of an electro mechanical foot controller, under which the driver placed the toe of his boot and to change gear simply "flicked" the pedal up, an operation that would be familiar to anyone who has ridden a motorcycle. Gear changes down were affected by toeing the pedal downwards at the same time as the right foot gave a quick burst of revolutions on the engine.

Located within the driving compartment were four vehicle batteries, two ammunition racks each side (holding five rounds apiece, which normally will be HESH) and all the normal instrument panels used for lighting and starting both the main engine and the auxiliary engine. An auxiliary engine was used to supply power to the vehicle electrical systems, especially when the vehicle was parked for a considerable period of time, allowing the main engine to be switched off (thus saving fuel and reducing smoke and noise).

The hull rode on six Horstmann suspension units bolted to the sides of the hull; by using this method overall height could be reduced compared to the use of torsion bars, which would have run across the hull and raised the profile of the vehicle. Each unit consisted of the bolt-on carrying bracket and two pairs of road wheels, one pair on each arm. Suspension was achieved by one arm moving up and compressing three coil springs retained by knife edge brackets on each axle arm. This is a similar system to that used on Centurion and Conqueror, was very simple and robust and it was relatively easy to re-place a damaged unit. Each station also carried a top roller, which supported the run of the track. The front Horstmann units were equipped with hydraulic shock absorbers, and although it was originally planned to have shock absorbers on the rear units as well, this was only ever applied to the prototypes due to weight.

The final drives were located at the rear of the hull, re-ducing the ratio of the drive from the gearbox to the actual sprockets, gearing the power to a sensible level. The final drives were bolted to the hull sides and could be replaced in the field if required. The rear hull plate held the armoured exhaust box which had three exhaust pipes, two were large bore for the L60 main engine and were located one on each side, a third smaller pipe was located on the right side above the main engine exhaust and served the auxiliary generator engine. Also on the exhaust box was the mounting for the gun crutch, used to retain the gun in the rear position for transport by road, rail or in the event of failure of the gun's powered stabilisation equipment. Later marks had the gun crutch replaced by a gun clamp of better design. On the left of the exhaust box the Infantry/tank telephone was mounted, which contained a telephone handset on a spring loaded cable. The idea that an infantry section commander could use the handset to contact the tank commander (to give him informa-tion or request some fire support) dated from Second World War experience. I can only recall this ever being used once in all my time on Chieftain, as the infantry just did not trust the tanks not to suddenly move off with them still gripping the handset, or worse still, to reverse!!!!

Located centrally on the hull is the turret and again this was a departure to normal practise, gone was the large armoured mantlet that was common up till then, in its place was a very narrow aperture for the 120 mm gun, this layout present a very narrow target front to the enemy and also reduced weight. The front half of the turret was again one single casting whilst the rear was of rolled and welded armour plate. The turret from the prototype and the Mk.1 although similar in appearance were actually very different. The turret housed the three remaining crew members, there being the gunner who was located on the right of the gun and below the commander who sat above him and on the left of the gun resided the loader.

The gunner had controls that allowed him to control and fire the weapons, on the floor he had two foot pedals; one operated the 7.62 GPMG and the other the .50 Rang-ing Gun. In front of him he had the elevating hand wheel which gave him hand control over elevation and depression

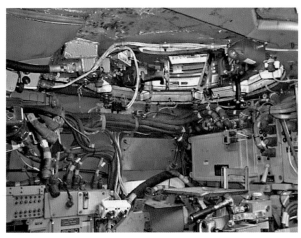

The commander's station in a Chieftain tank in the process of be-ing stripped out. Behind the commanders seat are two grey boxes, these contain all the equipment for providing power traverse and elevation. To the left of the commander's seat the black handle is his hydraulic pump that enables him to raise the seat till he can sit with just his shoulders above the cupola. This particular Chieftain was the authors very first command and now resides on Bovington Heath, all hatches now welded shut and is used by trainees for putting up camouflage nets [R. Griffin].

of the weapons, this coupled with a two speed hand traverse handle to his right were the controls most used for static firing. To control the gun while it was under power of the stabiliser he had a duplex controller (replaced in later marks with a thumb controller); this was a combined elevation and power controller all in one handle. Controlled by the gun-ners right hand, to fire the weapons whilst using the duplex controller he had a fixed grip with a firing switch located directly in front of his lap.

For sighting the gunner was equipped with a Sight Peri-scopic AFV No. 32 Mk.2 for his main sight and as a backup a sight unit (telescope No. 26) this was carried on through all marks (but was always notoriously inaccurate and difficult to keep bore sighted). The main sight could be swapped for an Infra-Red sight to enable night shooting, but carrying out this "swap" was always an interesting exercise between the gunner and the commander. The Infra-Red sight was used in conjunction with the 2KW searchlight mounted on the left of the turret, which could produce either white light or Infra-Red, and it was controlled by the Commander from his station. The gunner was also equipped with various control boxes to control lighting, sight washer/wipers and the turret power traverse and gun stabilisation.

Located on the left of the gun was the loader who was tasked with many responsibilities beyond loading the gun. He was responsible for loading the 120mm and the co-axial weapons, and the radio equipment but also for carrying out his "most important task" of making the good old English tea, using the Boiling Vessel or as it was better known the "BV". This was a very simple electric kettle and cooker rolled into one, nothing more than a square insulated box with a heating element within its double walls, and food could be cooked within in cans or boil in the bag sachets, and if you were naughty you could also cook chips although it was forbidden. The loader was responsible for clearing misfires (120mm) and

A Chieftain Mk.6 from the late Jacques Littlefield collection in America. This version has been fitted with the Pearson full width mine plough, all vehicles in this famous collection are fully operational, and although primarily a private collection it is viewed by students [R. Griffin].

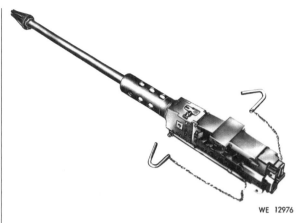

WE 12976

About the only illustration that can be found for the ill-fated American M85 ranging Gun, which was discarded as it was not capable of firing the UK ammunition without several breech explosions, this led to the introduction of the .50 Browning in the Ranging Gun role [Internet].

stoppages (7.62 mm). If all this was not enough he also was responsible for ensuring that the radios were working and on the correct frequency, and helped the commander take down orders and carry out decoding of radio orders. The loader also had a single periscope mounted forward of his location and he used this to help survey the battlefield for targets and threats (amazingly enough I have had my loader spot a target before either myself or the gunner picked them up).

The final crew man was the vehicle commander who could be either an officer, a Sergeant or a Corporal, ranks as a general rule for a three tank troop, but each Regiment had their own ways of doing it. The commander's job was to ensure the safety of the tank and its crew, to designate targets and control the gunner and loader during engagements by issuing fire orders (which would tell them what type of ammunition to load, what range and also what type of target they were engaging).

This is a shot of one of only two Mk.4 Chieftains ever built and its 4th/7th RDG crew in front of it. If you look at the rear instead of long bins the side has been continued up to allow greater fuel capacity, this and other ideas were incorporated from feedback form the Israeli experience of Chieftain. These hot weather trials took place at the Arizona Proving Ground in the USA. The fate of the two tanks is also rather different; one was destroyed by being used in mine trials, while the other has its turret removed and a Coles heavy duty crane fitted instead, in this role it was employed at the ranges at Kirkcudbright it then seems to have vanished [Ch. Trigg].

He would report all contacts to HQ and be responsible for working out all orders sent over the radio, he would also be in charge of all map reading and ensuring he knew the vehicle location at all times. Most importantly, he had to know how to work with his fellow commanders within his troop and squadron (Royal Armoured Corps equivalents to Platoon and Company) as a team. Tanks do not usually work alone, and engagements can only be won by making sound tactical use of other elements within a unit, no matter how small. At the end of a day the tank commander was (and remains to this day) the most mentally drained member of the crew.

The rear of the hull housed the power pack (the engine, auxiliary generating engine and transmission). In the centre of the engine deck a large "T" shaped platform sat directly above the power back offering access covers for checking and topping up the coolant and hydraulic oil. Either side were the main engine decks panels which lifted outward to the sides of the tank. Under these covers two radiators were located, once raised to the vertical position these allowed access to the L60 engine, auxiliary generator and air cleaners. At the very rear of the engine compartment the transmission was situated under its own set of deck panels, with the steering brakes on each side. There were oil and fluid levels to be checked daily in the transmission compartment as well. The entire power pack could be raised and removed in one unit for quick changes. One favourite trick on a cold day was for the crew to raise the transmission decks and sit on them with their feet on the hot components, although they had to be constantly aware of not resting too long otherwise their boot soles would melt. This related back once again to the lack of a crew compartment heater.

This was the layout for the Mk.1 Chieftain, which remained basically the same right up to the Mk.11 with many detail changes; some components were changed and most were improved. The frontal armour was 190mm on the glacis and 250 mm on the turret front, which would be supplemented on later builds, though actual thicknesses remain secret. With the Chieftain's excellent frontal ballistic shape the equivalent armour protection approached 400mm for the Mk.1. The basic gun tank layout evolved logically in detail to meet the needs mechanical improvements, fire control and gunnery improvements, armour and stowage layout improvements over 28 years of service.

So finally in 1966 the British Army had its new tank but it was accepted that the Mk.1 was woefully underpowered, and had other faults which needed to be rectified before the Chieftain could be issued as a potent front line vehicle MBT. The required improvements were quickly incorporated into the first production version of the Chieftain, the Mk.2. This was issued in November 1966 to the 11th Hussars (now the Kings Royal Hussars) who had just turned in their well-loved Centurions. The 11th Hussar's views on the new tank were mixed to say the least, but the Chieftains were with them, whether they liked it or not. The second regiment to convert was the 17/21st Lancers in early 1967, and the RAC units in BAOR all followed suit as deliveries allowed.

The Mk.2 Chieftain at first glance resembled the Mk.1, but closer inspection would reveal some big external differences. The No. 11 cupola had been replaced with a far superior version, the No. 15 Mk.1, the split hatches of the No. 11 had been replaced by a single commander's hatch that opened to the rear

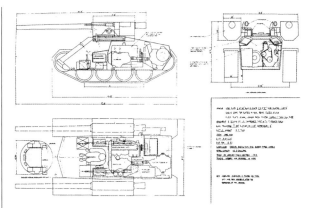

One of the surviving drawings showing the general layout of Project Prodigal, the miniature tank some thought would have made an economic Centurion replacement as a candidate for Medium Gun Tank Mk.2. Luckily sensible heads prevailed for once, and Prodigal was consigned to the history books, though almost nothing has been revealed about the project. As many as 3 prototypes may have been built in single or two-man layouts, but just by looking at the design one can see how overworked the single crewman would have been in this example [National Archives].

of the cupola. The No 15 cupola used 9 separate No. 40 Mk.1 periscopes fitted into the cupola, although on early Mk.2s some cupolas received No. 34 split periscopes. The No. 34 periscopes were a nightmare, as the crew had to fit one part from outside and another from inside, they never sealed properly and were the cause of many irritating leaks onto the commander's knees (the integrity of the tank's NBC seal must also have been compromised, and they were luckily speedily replaced). The view from the new No. 15 cupola was far superior to that possible from the No. 11, also providing the commander with a new sight, the No. 37 Mk.1 (which would evolve to the No. 37 Mk.6 during the Chieftain's lifetime). The No. 37 sight gave the commander X10 and X1 magnifications, the X1 used for general scanning and the X10 for target designation. The L60 Mk.5A engine, rated at 650 BHP was fitted; a small improvement on the Mk.1 but still woefully underpowered. The turret and glacis armour was improved (thickened, although specific changes

This view of the gun crutch as fitted to the early Chieftains is fitted to a Mk.1 preserved at the Vehicle Depot Ashchurch, Gloucestershire, England [R Griffin].

are still classified), although the suspension was still the same type as on the Mk.1 it had been strengthened and the removal of the rear unit's shock absorbers was carried over from the Mk.1. One very distinctive feature introduced on the Mk.2 (and only used on the Mk.2) was the full width splash screen fitted between the headlights. It was found that the driver needed this in winter, as it had been found that when driving through frozen water holes ice had a habit of riding up the glacis plate, and ran the risk of causing serious injury.

Some concerns from the trials in 1964 had still had not been addressed and would remain like that for a few more years. These mainly concerned the stowage, so whilst Chieftain had plenty of stowage bins these were taken up with the vehicle tools, leaving very little space for the crew's personal kit. The original turret basket only just had enough room for basic kit but also had to accommodate oil cans. Some Mk.2 crews tried to get around this, and I know that I did by using salvaged Centurion baskets from range targets fixed across the rear of the turret. While this was officially prohibited, many a blind eye would be turned, and photos of Mk.2s in the 1960s and 70s with M47 turret bustle bins fitted on the engine decks and even in place of the water can rack on the right hand turret side are well enough known. The other problem was with

the rear hull stowage bins, which were very large and sat on sloping rear mudguards with very little clearance between the bottom lip and the track. The consequence was that any large branches caught in the tracks would rip the rear mudguard off and damage the bins as the tracks dragged them around. The rear stowage bins would be replaced with a better designed pattern in due course but the money wasted must have awful in the early days of Chieftain's deployment.

The 120 mm L11 main armament on the Mk.2 remained the same as on the Mk.1, with the gunner having control over the power stabiliser system in very much the same way as for the Mk.1. The range was still determined by the use of the Ranging Gun firing controlled burst of three flashing tracer rounds; this was meant to be a controlled shoot with the burst going down very steady according to how the gunnery school taught it. The idea was to enable the commander to be able to determine which burst was which, but most crews usually got the burst off as fast as possible because in battle whoever gets the first round fired the quickest will usually win the battle. The ammunition load was 54 rounds of 120 mm which could either be loaded into the vehicle to a set standard load or specific to operations.

One other item of equipment fitted was the Infra Red detector stalk, which was located to the right rear of the

A superb view of one of the prototypes showing the final form before the redesigned hull for the production vehicles appeared. Points to note are the 120 mm cleaning rods strapped to the side, slope rear hull and lack of rear fuel caps, and also the No.11 cupola in the "umbrella" position, showing how it was meant to allow the commander to have his head out of cupola but with some measure of protection [R. Griffin].

commander's cupola, and its control box was located inside the turret by the commander's knee. In theory it was an excellent idea but in practice it was a dismal failure. The stalk had an Infra Red cover which protected three photo voltaic cells, which gave an overlapping cover of 360 degrees. Theoretically, once the system was set up if the tank was illuminated by an enemy using Infra Red the cells would detect this and trigger the alarm on the control box. The control box had three push buttons giving locations of front, left and right, and the commander would press each in turn; if the alarm was silenced by any given button then that was the area from which the Infra Red source had come... and if nothing happened with the three buttons then the signal was coming from the rear! The crew were then supposed to search for the enemy using their own Infra Red system and the two Kilowatt turret Infra Red searchlight and destroy them. The main problem was the sensors were too sensitive. The whole night fighting system was meant to be set up at last light, but as ambient light levels changed so the detectors kept setting the alarm off. In the end, especially if the Infra Red detection equipment was not specified for a given exercise, most troops left the stalk either in a tool bin or back in the troop stores at the barracks. More practical equipment would eventually be forthcoming for night combat.

Over 300 Chieftain Mk.2's were produced both at ROF Barnbow near Leeds and Vickers at Newcastle, and although built as original Mk.2's many of these vehicle eventually ended their lives as much higher mark vehicles, due to an extensive improvement program instigated by the Ministry of Defence called Exercise Totem Pole. As each modification program of the Totem Pole program was fitted to the vehicle its designation changed and at the end of this program in the late 1970s it was rare to see any Mk.2s in their original configuration, and almost none of these outside the UK. Some Mk.2s have survived as built, with a few in private ownership, and the rest as training aids (either as hard targets or being used to train REME recovery crews, where the vehicles are deliberately bogged down and then recovered).

This sort of upgrading was applied to most marks under similar programs, and it makes it difficult to confirm a mark visually because some marks were externally nearly identical; and some differences were strictly internal.

The next major build was the Mk.3, of which nearly 500 were built in 1969-71. The Mk.3 was also the first export version, the Mk.3/3P being built to Iranian specifications. The Mk.3 introduced many detail changes resulting from experience with the Mk.2 and ran to 12 sub-marks. The necessary modifications during the Mk.3's production indicate how

An example of how a vehicle mark would change:						
Build Mark	X	Y	Z	X&Y	X&Z	Y&Z	Final mark
2	2(X)	2(Y)	2(Z)	2(XY)	2(XZ)	2(YZ)	6

A rear view of the Mk.1 at Ashchurch vehicle depot this vehicle retains amazingly the orginal large rear bins although its rear wings have long gone, also notice the small bore of the exhaust pipes compared to production versions [R. Griffin].

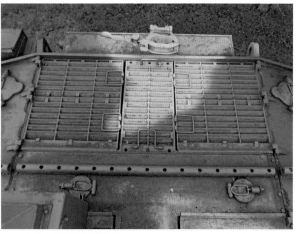

The original rear deck layout as seen on a Mk.1 Chieftain. Compare this to the shots of the prototype and the changes can be seen: horizontal rear deck and the two rear fuel caps, the two smaller caps are for the hydraulic fluid and coolant [R. Griffin].

much needed to be altered or improved, and it must be borne in mind that these modifications could not all be implemented at once. Among the more visible exterior modifications were some cosmetic changes. The large single headlights were replaced by twinned driving lights (with the outer pair being normal driving lights and the inner pair Infra Red lights), which led to the removal of the full width glacis splash board, and in its place a simple plate was fitted between the headlight units' brush guards. About ¾ of the way up the splash plate was angled forward to help deflect debris, and the plate was secured by two arms located on to the headlight brush guards.

At the rear of the vehicle one of them most striking changes to the layout involved the transmission decking panels; on the Mk.3 the entire hull rear had to be modified to accept the improved L60 Mk.6A power pack developing 700 BHP. This required a different exhaust system and to enable it to fit the rear transmission decks had to be raised several inches, achieved by inserting a new raised section and changing the transmission deck configuration (the early type on the Mk.2 had one central deck panel with two large panels either side, on the Mk.3 this changed to two small centre deck panels with one on either side now making four separate deck panels, which made lifting them much easier). This changed the rear deck layout from being flat to having a slight slope either side.

The commander received an improved cupola, the No. 15 Mk.2, which had got rid of all the locking clamps associated with the two part periscopes on the early No.15 Mk.1 cupolas. The rear of the turret also changed with the removal of the No 2 NBC pack and the fitting of the NBC pack No. 6, which ran the full width of the turret rear. This was a far superior product to the one it replaced. The gunner's sight was also modified for what was known as the extended range graticule. The introduction of this sight (known as the Sight No. 59) increased the ranging gun markings from 4 to 9, this along with the introduction of improved ammunition allowed the ranging gun to utilised out to longer ranges. Unlike the normal ranging technique which we have seen, anything from dot 5 upwards was controlled by the commander; also at this time the gunner's sight mounting was modified to accept the first of a range of Tank Laser Sights (or TLS). For a while both the TLS and

ranging gun were both fitted, and the gunner's Sight No. 59 had a special collar fitted to it so that it could slot easily into the TLS mounting. From a combat point of view the laser sight was probably the biggest advantage introduced with the Mk.3.

Externally some welcome changes to vehicle stowage were introduced on the Mk.3, and these included new, smaller rear hull stowage bins made of a much thicker metal, raised so they were horizontal instead of the sloping edge of the original mudguards. Unless the tank was reversed into something solid these mudguards lasted longer. On the turret the mesh stowage basket had been enlarged, and a smaller one was fitted on the commanders side as well, and this along with the introduction of a large commanders stowage bin (that replaced the jerry can holder on the right side of the turret) greatly improved crew stowage. The reliability of Chieftain's L60 engine began to improve, but never to the extent where crews had 100% confidence in it. The modifications that had appeared on the new production Mk.3s were eventually applied to the Mk.2s in service as they went through base overhauls during the mid 1970s.

Special mention must be made of an experimental mark that never went into service (although two vehicles were built),designated the Mk.4. Originally these were built in 1968 at ROF Leeds (Barnbow) to Israeli specification as part of the co-operation that existed between Israel and the UK at the time (Israel began to express interest in the Chieftain in 1964 and even after 1969; when a purchase was denied on political grounds, they made repeated approaches to the Foreign Office to purchase the tank). Israel was very keen to buy Chieftain to replace the Centurion on the grounds of their approval of British design philosophy based on combat use of the Centurion; and this would have been good news for the tank industry, but sadly politics reared its head and the UK cancelled the deal in 1969. The British government had loaned two Chieftain Mk.2s to Israel in 1967, and the Mk.4's specification reflected Israeli experience in operating tanks in hot climates. The Mk.4 would most likely have been the production version had the Israeli Chieftain order proceeded, but following the collapse of the attempts to secure an Israeli production license the Mk.4s reverted to British ownership.

It was decided that the two Mk.4s would be accepted into British service and sent on hot/dry trials in Arizona USA in May to September 1971. This was not the first time that Chieftain had been operated in that sort of climate though, for in 1966 1st Royal Tank Regiment had carried out trials of the Mk.2 in the desert area of Aden and had found that the original air cleaners were inadequate for the task. In the Mk.4 the test crews from 4/7th Royal Dragoon Guards and 13/18th Hussars were better provided for, and they found that in general the tanks performed very well. From the front, the Mk.4s looked like normal Chieftains, but the rear engine compartment had been raised to provide additional fuel storage, and in doing so the long stowage bins on either side of the rear hull were eliminated. The track fitted was notable for not having the rubber pads, and probably the most obvious feature on the tanks during the tests was the mass of instrumentation on the turret roof to monitor virtually every moment of the tanks working day. Eventually the trial came to an end and the crews and vehicles returned to the UK, which left the Ministry of Defence with two non-standard Chieftains, so one was used for anti tank mine trials and was eventually destroyed, though

This side view shows the T95 that was developed by the USA in the same time frame as Chieftain; the intention was to make the turrets interchangeable, although the logic for this was never very clear [M. Duplessis].

the fate of the second is not known and it might still survive, forgotten on a range or in storage.

The final new build Chieftain accepted by the British Army was the Mk.5 in 1972. All subsequent variants were based on either Mk.2, Mk.3 or Mk.5 rebuilds. Basically the Mk.5 was based on the Mk.3/3 but it also introduced its own new features. The Mk.7a L60 was fitted, but the gearbox had been strengthened for the 720 BHP it produced (giving a top road speed of 48 km/h). A new gunners telescope had been introduced (the No. 43) and the commander had the Mk.4 version of the commanders No. 37 sight. By this time Ranging Gun ammunition stowage had been reduced because the Tank Laser Sight had reached an acceptable level of robustness and reliability and within a few years the Ranging Gun would be removed completely. Main armament ammunition stowage was increased to 64 rounds by the inclusion of a ten round exterior projectile stowage box located in the left hand turret basket. This was only for APDS rounds, and the bin itself proved very unpopular because it took valuable stowage space away from the crew. It was often either used as crew stowage or taken out and left in the unit stores, only being fitted during range periods. The NBC pack was upgraded once again to the No. 6 Mk.2 and the L11 120 mm gun received a new lighter barrel, receiving the designation the L11A5, with a smaller and lighter fume extractor fitted and allowing the fitting of the Muzzle Reference System (MRS). This was a system to allow the gunner to confirm the relationship between the sight and the bore of the gun without dismounting; prior to this system the only chance to check this was when the gun was boresighted at the beginning of the day. Various factors could easily throw the gun/sight relationship out, such as hitting a tree or building, and obviously it was too dangerous and time consuming to check the bore after such an event, so the MRS was introduced to solve the problem and is used in a modernised form to this day.

The muzzle reference system consisted of a light source mounted on the turret, a mirror on the muzzle (covered with a shroud to protect it from the elements), a control box located near the gunner, a pair of level indicators (one fixed and one on the gun) and a modified gunners sight. If the gunner suspected a lost gun sight/relationship he would bring the gun to a pre-set position by means of the level indicators, move a lever on the sight (allowing the projected image from the mirror to be seen in the sight) and then switch on the light source. Light would be reflected from the mirror into the gunner's sight and would produce an image of the Bore sight mark, so the gunner could then adjust the marking in the sight so the two were superimposed and the gun/sight relationship would be corrected. He then would reset the lever and turn off the light source.

With the Mk.5's new L11A5 barrel a new thermal sleeve was also introduced, along with a gun clamp instead of the old gun crutch. The clamp was much easier to use when the gun had to be placed in it with gun rear as it had two arms which literally clamped the barrel, the old crutch had an arm that had to be forced over the barrel and locked, not always an easy task. There were also modifications to the commanders cupola although these were applied less uniformly, including a 90degree elevation for the commanders cupola 7.62mm GPMG, a fixed ammunition box instead of the box previously fitted to the machinegun cradle, and on some versions a large curved ammunition box on the left of the turret. These modifications were hoped to allow the crew to defend against helicopters or to engage infantry in upper floors of buildings.

Possibly the biggest step forward for the Chieftain Mk.5 involved the fire control system; specifically the introduction

T95 from the front showing a very narrow turret profile at the mantlet, on the left can be seen the lens of the optical range finder [M. Duplessis].

T95 on the range, although a new concept with its adjustable suspension the overall shape is very American, including the small sub turret that houses the commander and the inevitable .50 machine gun [M. Duplessis].

of the Improved Fire Control System (IFCS). As we have already observed the quickest shot is the one who will win the gunfight, and good as the Ranging Gun was it was 1950s technology; slow and liable to give away the firing vehicles location. To prevent this and to speed up engagement times the army introduced a computerised system that remained with Chieftain until the end of its service (fitted to the Mk.2 and Mk.3 series during rebuilds, and in a rather retrograde move a later version was also carried over to the Challenger 1).

At the heart of the IFCS system was an on-board computer located beneath the gunners seat, which received information from various sensors and data input by the gunner and commander. The commander and gunner both received new power controls which replaced the duplex controller with a state of the art single handle control fitted with a grip switch to engage the power system and to control traverse and elevation; a small thumb controller not unlike that found on some video game controllers today. A new Tank Laser Sight was fitted and into

this the computer would input an illuminated ellipse that would cover the target. Once all calculations were complete the computer would move to the correct aim-off mark and the gunner would then replace the ellipse back over the target and fire. To fire the gunner also had a new fixed firing handle resembling a fighter planes' joystick. Located on the top of the firing handle were a selector toggle switch and a safety cover. The gunner could select co-axial or main armament from this switch, and located on the left of the handle was the laser rangefinder/auto lay switch, which was a rocker type switch (one direction fired the laser and the other engaged the automatic gun laying stabiliser). This system was a great improvement on the Ranging Gun system but was not always as robust in the beginning, and as with all electronics of the era problems could arise, since this had been expected and to counter the loss of all IFCS systems the crew were taught what was known as *reversionary mode firing*. This entailed using the sight markings and going back to basic gunnery, known to the crews as "steam gunnery"; a reference to things old fashioned! The Mk.5 served as the definitive Chieftain, and as the basis for the Armoured Recovery Vehicle and Armoured Vehicle Launched Bridge. Just under 100 Mk.5 gun tanks were purchased for the British Army, although the type was used as the template to rebuild all Mk.2 and Mk.3 series vehicles in the Chieftain fleet. The Mk.5 was also selected as the basis for the Iranian and Kuwaiti armies' Chieftain orders (albeit in modified form) and as such was the most successful export variant. All subsequent British Chieftain marks would be rebuilds to Mk.5 standard or detail improvements of that standard to meet specific threats.

The increased power of the Mk.7A L60 led to severe reliability issues compounded by minimal engine running hours in the mid 1970s in an effort to minimise defence costs. By 1978 the problems with the L60 had caused political scandal and demanded a Parliamentary enquiry, which resulted in the definitive resolution of the L60's problems, a program known as Sundance. The Sundance modifications were applied to all Chieftains between 1978 and 1980, and resulted

A good rear shot of the T95 showing off the typical for the time domed cast turret, the commander small sub turret and .50 mg is also very clear in the shot [M. Duplessis].

in much-improved reliability. The Sundance modified engines received higher mark numbers, for example the L60 Mk.13A.

I can state with conviction that having had one of the very first Mk.13A packs fitted during a field training exercise in the late 1970s in my own Chieftain in the 4/7th Royal dragoon Guards, it did perform as expected and until the day we handed the Chieftain over to ordnance it never gave us problems. It was however, a 24 hour pack lift to fit in the field because some of the turret charge bins had to come out for re-routing of pipes and cables. When came the moment to fire it up for the first time, nothing... the Sundance power pack had come with faulty starters! Out it came again, new starters were fitted and it broke into life as sweet as anything. As we drove off we were told by REME (the Royal Electrical and Mechanical Engineers) "take it easy and run it in gently". We did go easy on the new engine for at least 5 miles, and then we opened it up and it never faltered. Sundance was a success because the government had no choice but to really deal with the issue because of the publicity, but it was an exercise that should not have taken place if things had been sorted out with proper development time in the first place.

Tribulations and Redemption: the 1970s and 1980s

The Mk.5 and earlier vehicles were consistently upgraded throughout the 1970s and 1980s to maintain the Chieftain as a potent weapon system. The first issue was to upgrade the initial production batches of Mk.2 and Mk.3 series vehicles. The Mk.6 was the next variant and it basically was a Mk.2 with all modifications complete to Mk.5 standard (and it ran to four sub versions; major changes were the final removal of the Ranging Gun, modifications for the stowage of APFSDS rounds. The Mk.6 also included provision of an Image Intensified sight for the commander- although it was rarely seen in service, if at all. The "I.I." sight involved the commander removing the No. 37 day sight and fitting the L5A1 Image Intensified sight, and even if issued I do not think it would have been used much as the principle was a step back to Mk.2 days when both commander and gunner both had to swap day sights for night sights at last light. This was not a popular task and created problems with loss of gun/sight relationship. The Mk.6 was also modified to accept a training aid called SIMFICs, this was a weapons effect simulator that had a laser mounted in the 120mm bore connected to the sights, laser sensors, sound effects and finally a box to simulate the weapons firing mounted around the turret. It was a very complex system that took a lot of time to set up and never worked 100% on all the vehicles, but it was a useful training aid. The Mk.7 was a Mk.3 series tank with all its modifications complete to the same standard as a Mk.6, with no further major changes included in the upgrade. The Chieftain Mk.8 was a minor upgrade of the Mk.5 respectively to the latest refit standard and again no major changes were included. When the Mk.6,7 and 8 were overhauled to modify the main armament projectile stowage for APFSDS (Armour Piercing Fin-Stabilised Discarding Sabot) munitions, the designation changed to Mk.9.

The next big modification, executed on the Mk.9 embodied two major changes, firstly the NBC pack was upgrade to the No. 11 Mk.1, which looked very much like the No.

S086

Head on view of a prototype Chieftain showing the false bin over the glacis looking rather the worse for wear, this vehicle still has the wing mirrors fitted, in reality although issued up till Mk.11 they were rarely fitted and the driver relied on the commander for instructions on the main road [R. Griffin].

6 NBC pack externally, but the biggest upgrade was to the vehicle's frontal armour. This changed the designation of all tanks so modified to Chieftain Mk.10. The turret front and glacis plate around the turret ring received a heavy appliqué layer, drastically changing the turret profile (in my opinion making the tank look better!). This armour was added in the 1982-84 period to improve frontal immunity to ATGMs like the Malyutka/Sagger and to counter the threat posed by the 125 mm guns fitted to MBTs like the T-72 and T-64. As a result of intelligence from the Iran/Iraq war, it was found that Iraqi T-72 125mm rounds could penetrated the conventional frontal turret armour of Iranian Chieftains at ranges of 1 km. As insurance that the Chieftain could withstand 125 mm fire at normal combat ranges, the Stillbrew armour package was devised and fitted to nearly all BAOR Chieftains during base rebuilds. At the time rumours abounded about the composition of Stillbrew, and the lurid tales of what it was made of ranged from armour steel to the latest version of Dorchester (Chobham) armour, but the truth was far more mundane. Stillbrew was nothing more than a thick layer of armour plate backed by several layers of rubber. It functioned as stand-off armour and was a cheap expedient that reduced the effectiveness of an incoming Armour Piercing projectile while also offering increased protection from shaped charge warheads. While trial versions of the Chieftain (which will be treated in the elsewhere) had included the FV4211, replete with Chobham armour; as early as 1969-70, the cost of such a comprehensive modernization of the Chieftain's protection was accepted as prohibitive. Stillbrew armour was fitted on the turret front and on either side of the drivers hatch in order to protect the turret ring. The Mk.10 version and the Mk.11 which followed it were the most common versions found operationally at the end of Chieftains life, and these served alongside the Challenger MBT in the RAC units of BAOR and in the UK into the early 1990s.

The Mk.11 was the last in line and was identical to the Mk.10 apart from one major change which significantly improved the type's night fighting capability. Gone at last was the now obsolescent Infra Red searchlight and in its place TOGS (Thermal Observation and Gunnery System) was installed. The TOGS system was a fantastic piece of equipment and by the mid 1980s was long overdue. It was located in a lightly armoured barbette similar to that fitted on Challenger 1 but located on the left side of the turret. Both commander and gunner had their own small screen monitor in one of the downsides of the equipment it was "added on" (instead of being inherent to Chieftain's original design). Components had to be fitted in the best available locations rather than in the ideal locations, but that apart TOGS is a battle winning piece of equipment. Sadly not all the Chieftain Mk.10s were converted to TOGS, due both to cost and the impending decision taken to retire the Chieftain and buy a new main battle tank to serve alongside the Challenger 1 in the 1990s. A proposed Chieftain Mk.12 was designed as a rebuild to supplement the Challenger, but never was built, if it had been it would have been a Mk.11 fitted with the L30 120 mm high pressure gun as fitted to Challenger 2 along with other modifications. The final export production model of Chieftain was in fact the Mk.15 but this was the designation given to tanks supplied to Oman, built to Mk.9 standard.

So we end the Chieftain story; much maligned for its engine in its time, it is now recalled with great fondness (and even has its own Facebook page). No Chieftains serve with the British Army now, the last gun tanks retired from 1st Royal tank regiment in 1994, the AVLB and AVRE were the last to leave and ironically were the only British Chieftains to go to war during the Operation Granby. Chieftain was designed to truly fight on the nuclear battlefield; a design that was heavier, better armed and better armoured than its contemporaries within NATO and its potential opponents in the Warsaw Pact. It retained excellent battlefield mobility when it was in good mechanical order. It was by far the most powerful tank in any army at the time of its introduction and remained so until the Leopard 2 and M1 Abrams became available in 1979. It subsequently was an excellent and capable second to the Chobham-armoured Challenger 1 in BAOR. It was the last conventionally armoured MBT fielded by the RAC and was in many ways advanced for its time. Though it would see no combat in British hands and would only see battle in the hands of disadvantaged Middle Eastern armies bereft of the level of infrastructure it cer-

An unusual shot showing an area not usually visible, this is the gear that the turret drive motor engages to traverse the gun, on top of this gear is located the turret ring which is nothing more than a very large bearing which supports the turret [R. Griffin].

tainly needed (certainly in terms of its mechanical fragility), there is every reason to believe that it would have taken a heavy toll of its attackers had it been used (as designed) to defend the frontage of the 1st British Corps in the Northern Army Group of NATO. In all fairness the western MBTs of the first generation, with the exception of the Leopard 1 and the M60 series, were little better in terms of mechanical reliability. Given the T-72's vulnerability to the L7 105 mm gun it is certain that the L11 120 mm gun would have proven even more punishing to even the most modern Soviets tanks of the period, at ranges beyond the 125 mm weapon's best performance. Although upgrade packages were studied to further strengthen the Chieftain's protection with Chobham armour, it was never employed on the BAOR Chieftain fleet; a relevant point when one considers that the Chieftain's main nemesis, had war come in the 1970s, would probably have been the Sagger ATGM, and not enemy MBTs. Like many British weapons of the post-war period, the Chieftain was a missed opportunity in some ways, but it is well remembered for its 28 year long watch.

Chieftain Main Battle Tank. Development And Active Service From Prototype To Mk.11 • Robert Griffin
First edition • LUBLIN 2013 • ISBN 978-83-62878-52-9

© All rights reserved. With the exception of quoting brief passages for the purposes of review, no part of this publication may be reproduced without prior written permission from the Publisher.

Book editors: **Merlin Robinson, Tomasz Basarabowicz, Stanisław Powała-Niedźwiecki** • Color profiles: **Sławomir Zajączkowski** • Cover photo: **Plain Military** • Photos: **S. Jacklin, Fort Knox, J. Hall, R. Griffin, Plain Military, P. Rogers, Ch. Trigg, M. Duplessis, R. Lamb, 4/7 RDG magazine, P. Hoskins, MoD, REME, RTR, RH, P. Richards, KRH, HQ BATUS, QRIH, P. Cook, T. Walker, P. Breakspear, P. Holder, B. Grundy, Tank Museum - Brussels, Kirkcudbright Range Control** • Scale drawings: **Jarosław Dzierżawski** • Design: **KAGERO STUDIO, Łukasz Maj**

Oficyna Wydawnicza KAGERO • www.kagero.pl • e-mail: kagero@kagero.pl, marketing@kagero.pl

Editorial office, Marketing, Distribution: KAGERO PUBLISHING Sp. z o.o.,
ul. Akacjowa 100, os. Borek, Turka, 20-258 Lublin 62, Poland, phone/fax (+48) 81 501 21 05

A Mk.10 Chieftain fitted with the SIMFICS weapons simulator, notice the detectors around the turret and the "flash bang" projector located centre of the turret. The crew have fitted half a Chieftain long stowage bin to the splash plate to increase their stowage [R. Griffin].

A Mk.10 Chieftain in its natural element on the ranges along with its partner vehicle, the FV432, with which it was paired during its long career in BAOR. The FV432 outlasted the Chieftain in service by about 15 years [Plain Military].

A refuelling scene probably on SOLTAU training area West Germany. Notice that all the troops are full dressed in their chemical warfare suits including respirators [J. Hall].

This is a KAPE (Keep The Army In The Public Eye) posing with their Chieftain Mk. 2 somewhere in N Yorkshire, of note is the full width splash guard which besides serving for its original purpose was also very handy for stowing the drivers kit behind [4/7 RDG

A troop of Chieftain Mk.10's lined up during a demonstration at Bovington camp, probably in the late 1980s. Of interest is the use of live foliage for camouflage, something that in these environmental and politically correct days would not be allowed. [R Griffin].

A Mk.2 in a typical scene on Soltau training area, very visible are the large old style fume extractor and the full width splash plate. This vehicle is also fitted with Simfire the predecessor to Simfics. The laser can be seen on top of the barrel near the turret while behind it is the flash/bang generator [R. Griffin].

Nice head on view of Mk.2 fitted its splash shield, the cam net draped across the turret as in the picture was not the best way of carrying it, as it tended to obscure the gunners sight, it more often was turned around so the long edge ran across the turret rear [P. Hoskins].

This shows just what can be achieved in the line of field repairs, two wheeled recovery vehicles have managed to lift the complete Chieftain turret away from the hull. Although this was a staged publicity shot it could have been done for real if required [REME].

One tank in each Squadron was fitted with the dozer blade and this was usually the 2 i/c. The blade was an unpopular fitting as very often it failed, the power pack for it meant losing the front right stowage bin and it was over complicated compared to later versions. It really came into its own though during the winter season in BAOR when it was used to clear the roads in camp [R. Griffin].

A fairly standard shot of a Mk.5 Chieftain? It has no smoke dischargers! In their place are the 12 boxes of the British Army's proposed smoke discharger replacement known as V.I.R.S.S. (Visual Infra-red Screening System). Much publicity was made of this system and then it was dropped. The tank also has TOGS fitted to the left side of the turret [MoD].

Whilst not a pretty sight it does show how a tank crew live, notice how the bazooka plate has been lowered to the horizontal position thus forming a handy table, also the home comforts of a transistor radio on the tank, in front of the crew man are some of the famous compo rations in a tin [RH].

To save wear and tear on the tanks during long road marches the will be carried by tank transporters and this shows a Chieftai training aid being carried onto Salisbury plain by the Britis Army's latest transporter, the Oshkosh, American built but t British requirements, it replaced the Scamell Commander. Thes vehicles when moving through a town take on gigantic propor tions but are actually very manoeuvrable [Plain Military].

A similar shot but close up showing just how worn out these Chieftain hulls are and the size of the load on the transporter [Plain Military].

An all too familiar view of Chieftain, with one of its major assembly's being removed, in this case the David Brown TN12 gearbox, on the floor behind th tank can be seen L60 power packs that most likely are awaiting repair, at least th crew have the satisfaction of an overhead hoist and working indoors [P. Richards]

The depressing view once all has been removed, notice the nice sheen of oil lying on the hull floor, under the cross frame can be seen were an access plate has been removed, this is the drivers job and usually involved him getting a good soaking in all the fluids from the assembly's that had gathered on the hull floor. It had long been a request from crews for a drain plug to be fitted to allow the worst to be removed before taking the plate off, however MoD could see no reason for it and refused. It was incorporated but much later in Chieftains life [P. Richards].

The Chieftain Mk.1 and Mk.2 carried this jerrycan rack on the right hand side of the turret in the location later used for the commander's stowage bin on later marks. It held 2 jerrycans (for water of course!) and the box beneath contained smoke grenades for the smoke dischargers fitted to the turret on each side of the main armament. The lack of stowage arrangements was seen to be a serious problem on the Chieftain Mk.1 and Mk.2, which lead to the use of derelict M47 turret bustle bins being fitted instead of the jerrycan rack by some crews for exercises. The use of wire mesh baskets on both sides of the turret bustle was introduced at around the same time as the rear hull stowage bins were reduced in size. On the Mk.3 this problem was addressed with the introduction of the commanders stowage bin [RTR].

The No. 11 Cupola fitted to the prototypes and Mk.1 Chieftain was inspired by the commanders hatch arrangement seen on the Centurion Mk.8 and Mk.10. The vehicle shown is a Mk.1 preserved at the Vehicle Depot, Ashchurch, Gloucestershire, England [REME].

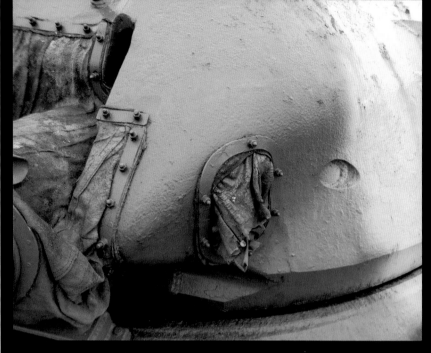

This view shows the canvas cover that would have protected some of the RG barrel where it protruded from the turret, to the left can be seen the mantlet cover fitted to the 120 mm, this was designed to prevent dirt and dust entering the turret [R. Griffin].

This shot show the No.2 NBC system fitted to early marks of Chieftain, air was drawn through the top and through and not to efficient single filter and then ducted to the turret via the trunking seen to the left of the unit [R. Griffin].

Tank crews could make themselves very comfortable on the back decks to take advantage of the heat from the engine. They would put the gun rear, fully elevate and then drape the tank sheet which is a large plastic cover over the barrel, tie it down and add more canvas as required as in the photo; also you can see the use of the 120mm cleaning rod acting as tent poles [KRH].

A beautifully restored Mk.11 Chieftain marked in the scheme found at BATUS, notice the Thermal camera inside the barbette [R. Griffin].

This is a Chieftain Mk.10 on the prairie at BATUS in Canada. Of special note is the white line running across the turret, a similar line is on the opposite side of the turret and they are known as the "45's" if there is nothing within the view of both lines and the tank has been given the all clear then it may fire [R. Griffin].

Beautifully preserved Mk.11 located outside the headquarters building at BATUS Canada, this vehicle finished its time in Canada and was refurbished for duty as a gate guard, making it one of the lucky ones [Dan Hay].

The second tank in Squadron Headquarters makes its way across the BATUS prairie, it has a well-worn look to it, notice in front of the driver is the desert camouflage net rolled up and stowed behind the splash guard. Along the bazooka plates are tied poles to be used in erecting the camouflage net as the idea of the net is to keep it away from the vehicle, this helps break up the shape and also makes it much easier for the crew to move around under it [R. Griffin].

A typical scene from BATUS with the L60 removed and either waiting for a new pack or it will be worked on there. Or interest is the varying modes of dress; the REME soldier on the right is wearing his NBC suit, minus gloves, boots and mask, the Queens Royal Irish Soldier leaning on the L60 has his normal green overalls underneath a DPM combat jacket, the crew man working on the road wheels is wearing a one piece quilted suit, known to the troops as a "Chinese fighting suit" due to is resemblance to the uniforms worn by the Chinese in cold weather [QRIH].

he Berlin urban camouflage scheme is shown off in this shot ken in Cornwall; there were colour variations to the scheme is showing lighter colours than normal [R. Griffin].

Fine head on view of a Mk.3 painted in the urban camouflage developed by D Squadron 4th/7th Royal Dragoon Guards during their tour of Berlin. The canvas cover that would protect the muzzle of the RG is shown hanging down indicating that no RG is fitted [P. Cook].

superb shot of the Berlin squadron parading on Allied forces day in Berlin unusually for British camouflage it has been applied ith care and following a set pattern [R. Griffin].

These 5 photos show a Chieftain Mk.10 on display at the Imperial War Museum Duxford in 1999. The vehicle in question is a rather weathered ex-Berlin Brigade Armoured Squadron tank replete with the markings of the 14/20th Hussars. Note how the engine decks are covered with mesh panels, typical for late model Chieftains. The rear convoy safety markings, while faded in this case, would perhaps have been less of an impediment to camouflage in the urban environment in the British sector of West Berlin than elsewhere in BAOR. The basic urban camouflage scheme was quite effective and endured from the early 1980s until the Berlin Garrison stood down [M.P. Robinson].

A rare shot of Chieftain in combat although it an Iranian model taken during the Iran/Iraq war. Notice the amount of clutter around and on the tank and how the crew have piled sandbags against the wheels [R. Griffin].

A tank park in the Middle East full of left over armoured vehicles from the Iran/Iraq war. Most have been scavenged probably to obtain parts to keep others running. The tank nearest to the camera has had its power pack removed and interestingly lying on the front right wing is its air cleaner, this is also painted sand colour whereas it normally is duck egg blue [Internet].

Taken at Bovington during the Annual Tank Fest, this is the Chieftain we should have had, it is in fact an ex Jordanian Khalid, which was developed from Shir 2 designed for the Shah of Iran before he was overthrown. It has a Rolls Royce CV12 diesel and new gearbox, improved suspension, commander is supplied with a day/night sight, and sadly the UK had to battle on with its standard Chieftains [T. Walker].

A very good restoration project for someone, one very battered Chieftain photographed in Iraq after Gulf War 1, this would be a vehicle that may have been captured from the Iranian and put to use by the Iraq Army [R. Griffin].

Chieftain Tank of the Kuwaiti Army after the first Gulf War, although not making the media Chieftain tanks of the 35th Armoured brigade engaged the invading Iraqi forces and inflicted quite an amount of damage, but at the end of the day they had run low on fuel and ammunition so had to withdraw to Saudi Arabia. They were also in action again during the war as part of the coalition forces. One officer is reported to have said "Chieftain is the best tanks in the world and if we could get it to where it is meant to be we can prove it" a dig at the unreliability of the L60 again [Internet].

This is one of Chieftains that was used by the Omani Army until recently, of interest is the camouflage colours, the different type of antenna base and the welded portion were the Ranging Gun used to be [P. Breakspear]

Order of battle and wireless call sings of the 4/7th Royal Dragoon Guards, Sennelager ,West Germany 1970

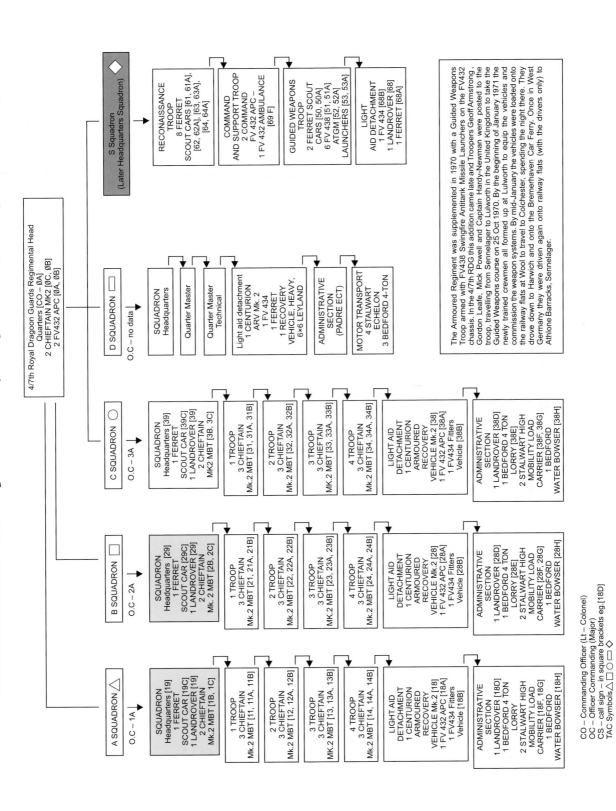

4/7th Royal Dragoon Guards Regimental Head Quarters [CO – ØA]
2 CHIEFTAIN MK2 [ØC, ØB]
2 FV432 APC [ØA, ØB]

A SQUADRON △
O.C – 1A

SQUADRON Headquarters [19]
1 FERRET
SCOUT CAR [19C]
1 LANDROVER [19]
2 CHIEFTAIN
Mk.2 MBT [1B, 1C]

1 TROOP
3 CHIEFTAIN
Mk.2 MBT [11, 11A, 11B]

2 TROOP
3 CHIEFTAIN
Mk.2 MBT [12, 12A, 12B]

3 TROOP
3 CHIEFTAIN
Mk.2 MBT [13, 13A, 13B]

4 TROOP
3 CHIEFTAIN
Mk.2 MBT [14, 14A, 14B]

LIGHT AID DETACHMENT
1 CENTURION ARMOURED RECOVERY VEHICLE Mk.2 [18]
1 FV 432 APC [18A]
1 FV434 Fitters Vehicle [18B]

ADMINISTRATIVE SECTION
1 LANDROVER [18D]
1 BEDFORD 4 TON LORRY
2 STALWART HIGH MOBILITY LOAD CARRIER [18F, 18G]
1 BEDFORD
WATER BOWSER [18H]

B SQUADRON □
O.C – 2A

SQUADRON Headquarters [29]
1 FERRET
SCOUT CAR [29C]
1 LANDROVER [29]
2 CHIEFTAIN
Mk. 2 MBT [2B, 2C]

1 TROOP
3 CHIEFTAIN
Mk.2 MBT [21, 21A, 21B]

2 TROOP
3 CHIEFTAIN
Mk.2 MBT [22, 22A, 22B]

3 TROOP
3 CHIEFTAIN
Mk.2 MBT [23, 23A, 23B]

4 TROOP
3 CHIEFTAIN
Mk.2 MBT [24, 24A, 24B]

LIGHT AID DETACHMENT
1 CENTURION ARMOURED RECOVERY VEHICLE Mk.2 [28]
1 FV 432 APC [28A]
1 FV434 Fitters Vehicle [28B]

ADMINISTRATIVE SECTION
1 LANDROVER [28D]
1 BEDFORD 4 TON LORRY [28E]
2 STALWART HIGH MOBILITY LOAD CARRIER [28F, 28G]
1 BEDFORD
WATER BOWSER [28H]

C SQUADRON ○
O.C – 3A

SQUADRON Headquarters [39]
1 FERRET
SCOUT CAR [39C]
1 LANDROVER [39]
2 CHIEFTAIN
MK2 MBT [3B, 3C]

1 TROOP
3 CHIEFTAIN
Mk.2 MBT [31, 31A, 31B]

2 TROOP
3 CHIEFTAIN
Mk.2 MBT [32, 32A, 32B]

3 TROOP
3 CHIEFTAIN
MK.2 MBT [33, 33A, 33B]

4 TROOP
3 CHIEFTAIN
Mk.2 MBT [34, 34A, 34B]

LIGHT AID DETACHMENT
1 CENTURION ARMOURED RECOVERY VEHICLE Mk.2 [38]
1 FV434 Fitters Vehicle [38B]

ADMINISTRATIVE SECTION
1 LANDROVER [38D]
1 BEDFORD 4 TON LORRY [38E]
2 STALWART HIGH MOBILITY LOAD CARRIER [38F, 38G]
1 BEDFORD
WATER BOWSER [38H]

D SQUADRON □
O.C – no data

SQUADRON Headquarters

Quarter Master

Quarter Master Technical

Light aid detachment
1 CENTURION ARV Mk. 2
1 FV 434
1 FERRET
1 RECOVERY VEHICLE, HEAVY, 6×6 LEYLAND

ADMINISTRATIVE SECTION
(PADRE ECT)

MOTOR TRANSPORT ECHELON
4 STALWART
3 BEDFORD 4-TON

S Squadron ◇
(Later Headquarters Squadron)

RECONAISSANCE TROOP
8 FERRET
SCOUT CARS [61, 61A], [62, 62A], [63, 63A], [64, 64A]

COMMAND AND SUPPORT TROOP
2 COMMAND FV 432 APC –
1 FV 432 AMBULANCE [69 F]

GUIDED WEAPONS TROOP
2 FERRET SCOUT CARS [50, 50A]
6 FV 438 [51, 51A]
ATGM [52, 52A]
LAUNCHERS [53, 53A]

LIGHT AID DETACHMENT
1 FV 434 [68B]
1 LANDROVER [68]
1 FERRET [68A]

The Armoured Regiment was supplemented in 1970 with a Guided Weapons Troop armed with FV438 Swingfire Antitank Missile Launchers on the FV432 chassis. In the 4/7th RDG this addition came late and Troopers Geoff Armstrong, Gordon Leafe, Mick Powell and Captain Hardy-Newman were posted to the troop, travelling from Sennelager to Lulworth in the United Kingdom to take the Guided Weapons course on 25 Oct 1970. By the beginning of January 1971 the newly trained crewmen all formed up at Lulworth to equip the vehicles and commission the weapon systems. By mid-January the vehicles were loaded onto the railway flats at Wool to travel to Colchester, spending the night there. They drove down to Harwich and onto the Bremerhaven Car Ferry. Once in West Germany they were driven again onto railway flats (with the drivers only) to Athlone Barracks, Sennelager.

CO – Commanding Officer (Lt – Colonel)
OC – Officer Commanding (Major)
CS – call sign – in square brackets eg.[18D]
TAC Symbols △ □ ○ □ ◇

Order of battle and wireless callsings of the 4/7th Royal Dragoon Guards, Detmold, West Germany 1989

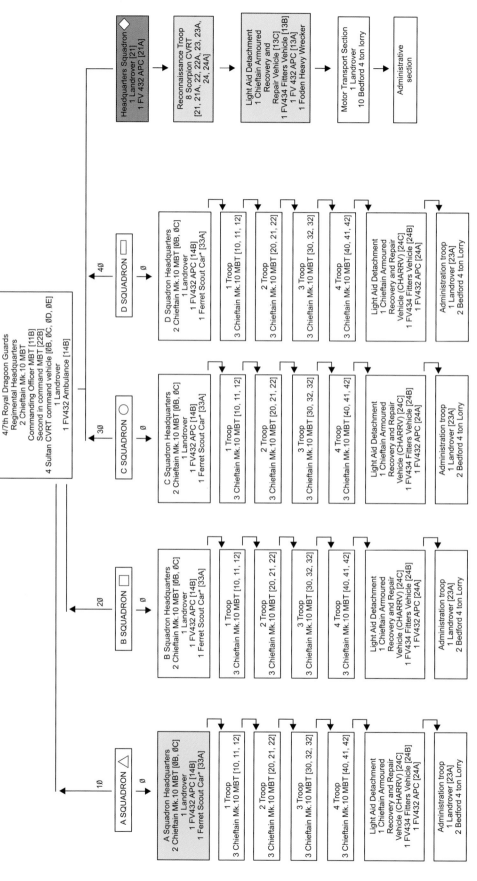

1. Each squadron preceded its call sign by a code letter, which was changed each day. Markings included squadron geometric symbols
2. By 1989 the Guided Weapons Troop had been removed from the Armoured Regiment and was reformed as a Brigade-Level Detachment based on CVRT Striker
* Squadron sergeant–major used a Ferret Scout Car which was replaced during 1929 a Spartan CVRT APC
CO – Commanding Officer (Lt – Colonel)
OC – Officer Commanding (Major)
CS – call sign – in square brackets eg.[18D]
Ø – 0 and was pointed on vehicles as Ø ⊚ ZERO
TAC Symbols △ ☐ ○ ☐ ◇

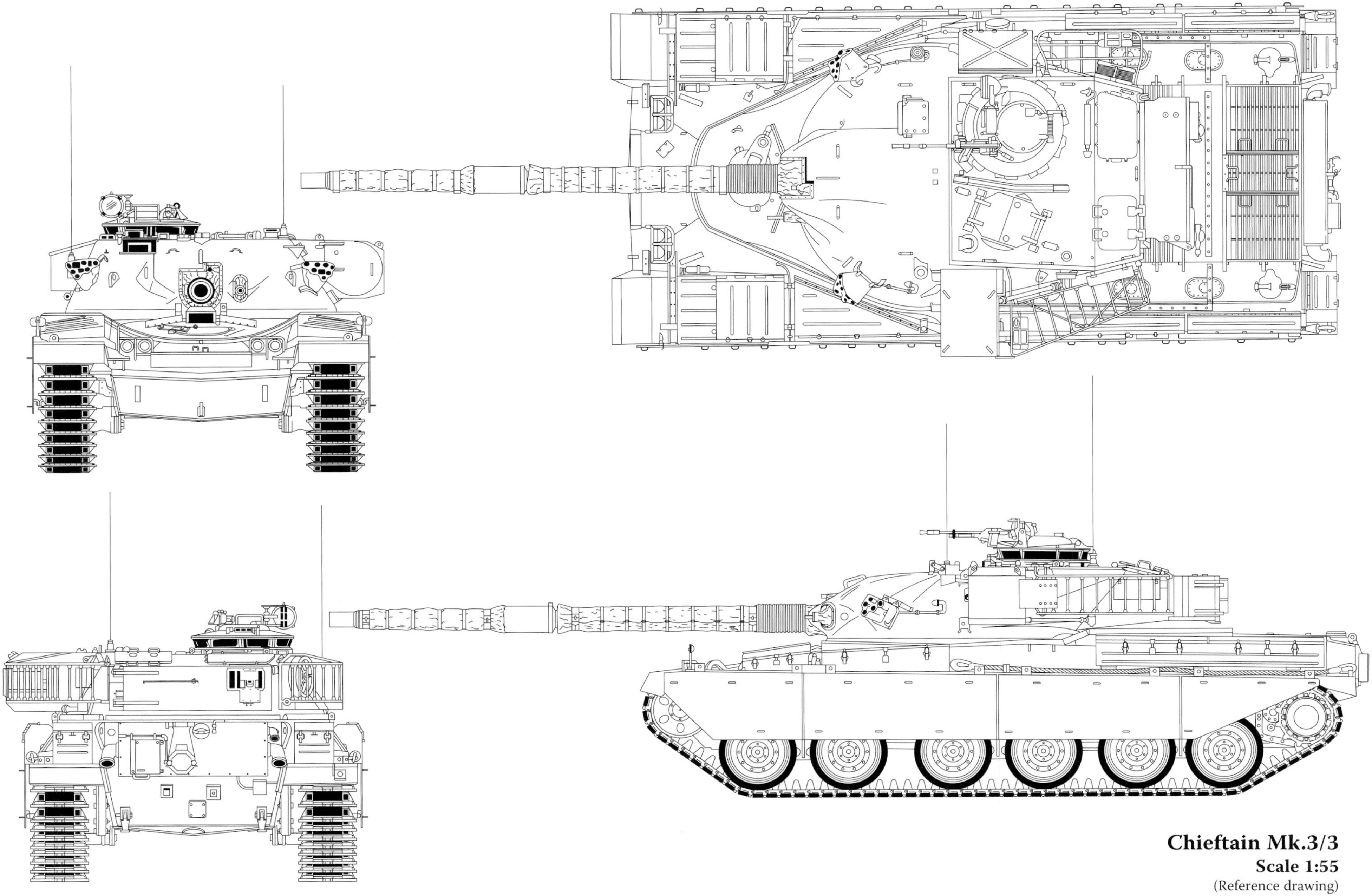

Chieftain Mk.3/3
Scale 1:55
(Reference drawing)

Drawings Jarosław Dzierżawski

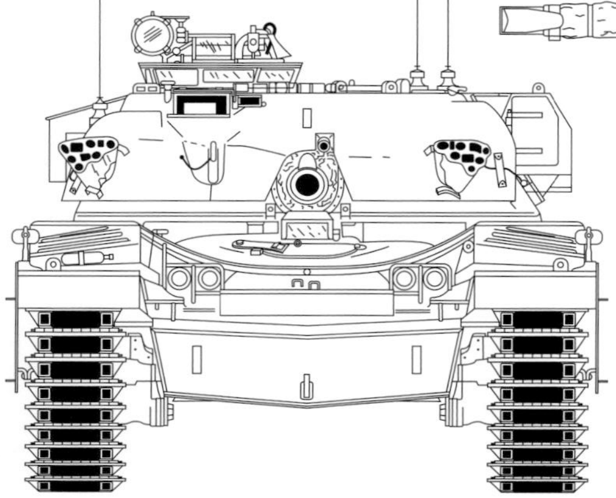

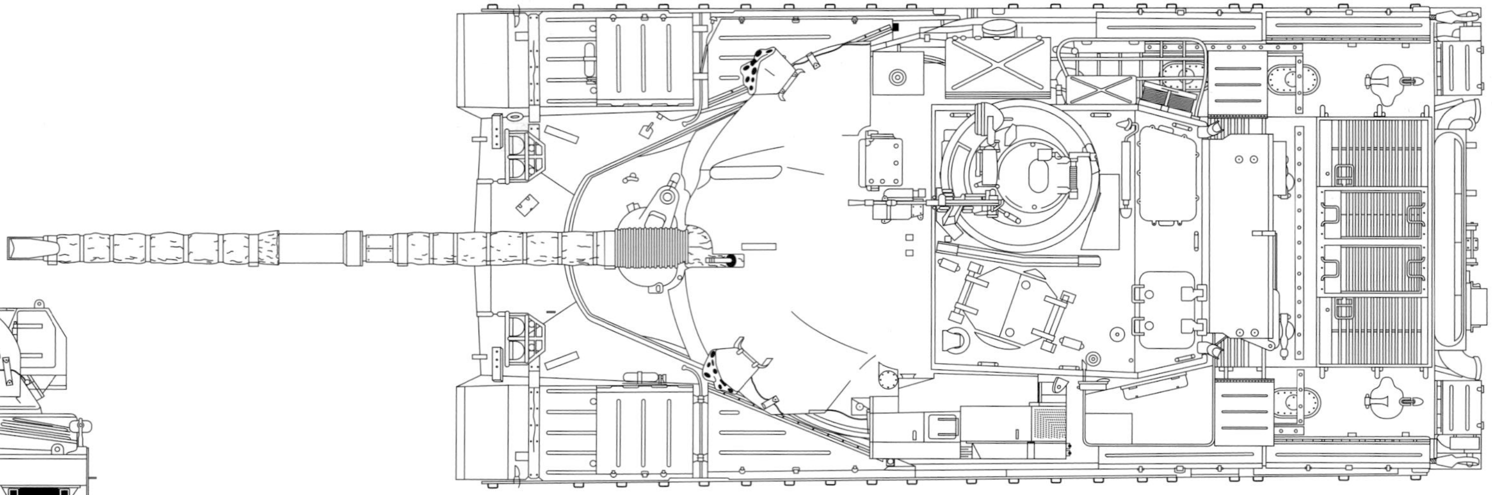

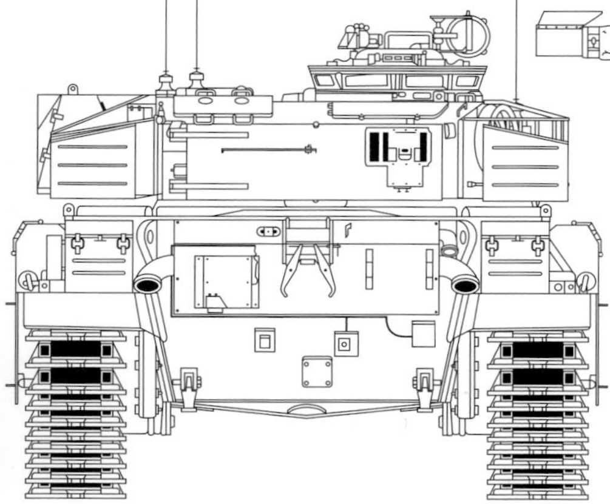

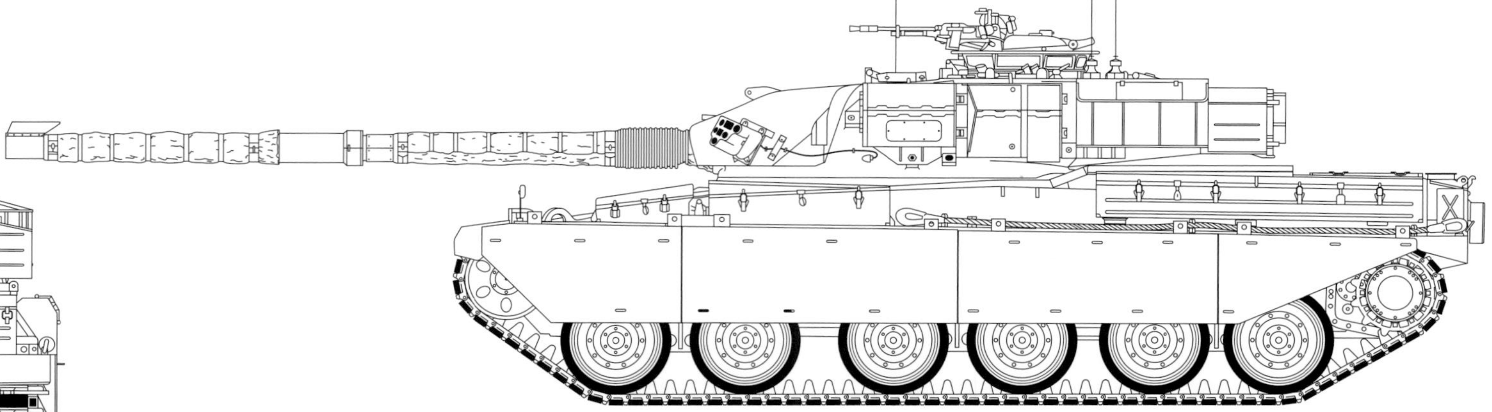

Chieftain Mk.11/C
Scale 1:55
(Reference drawing)
Drawings Jaroslaw Dzierżawski

A sad end for a Mk.2 Chieftain, situated on Warcop ranges as part of the conventional forces treaty Europe, centre suspension units had to be removed hence the forlorn look [R. Griffin].

Another Mk.2 waiting its fate, of interest is the fact that although a Mk.2 it does have the later headlight array and the L5 barrel [R. Griffin].

This was to be the fate of many Chieftains, soldiering on as hard targets or static pieces on the training area. This view shows the modified rear transmission decks and gun clamp and the extra two bins at the rear of the turret which distinguish this as a Mk.11 [R. Griffin].

A sad sight indeed but it is what became many Chieftains on re-tirement, this is located at the gunnery ranges at Lulworth Cove in Dorset. Although very badly battered it does give some idea as to the frontal armour thickness [R. Griffin].

This shows Cpl Ricky Lamb of the 4th/7th Royal Dragoon Guards checking his map while operating closed down. The shot does give and idea of the limited space available for the vehicle com-mander [R. Lamb].

Another lucky Mk.11 that is sitting its retirement out in the sunshine, notice the gun clamp as opposed to a gun crutch on the ex-haust box and to the right of the clamp the two sets of hooks, these are used to carry up to three spare track links to aid in self help repair [P. Holder].

Looking very much like a Troop of Chieftain lined up ready to move out but sadly these are sat waiting on Warcop ranges for use as hard targets, again notice the removal of the entre suspension units [S. Jacklin].

While this might appear to be a very well maintained in service vehicle it is in fact the result of a restoration by British Military Vehicles, who have performed some amazing restorations over the years [B. Grundy].

This fine example of a Mk.11 Chieftain is located outside Castlemartin ranges in South Wales, it is name Romulus and on the othe side of the entrance is a German Leopard called Remus, both name after the founding twins of Rome according to Legend. This tan replaces a Conqueror which itself is now going to be refurbished and saved [R. Griffin].

Although located indoors in a museum this show how the tank crew shelter known as "the bivvi is tied to the tank. This versatile shelter gives the crew protection from the elements and somewhere to sleep in relative comfort. Enterprising crews will think of many variations on how to erect th bivvi; it can be strung between two trees and ever arranged so that the front is propped up giving a sort of lean to effect [R. Griffin].

Here is a privately owned Chieftain working as it was meant to, owned by Jack Cross of Alberta, Canada. The vehicle is well looked after and regularly comes out to play! [J.Cross]

The ultimate road load 100 tons of Commander Transporter and Chieftain Tank, the tank is a Mk.2 and they are on show at a military show somewhere in the UK [Plain Military].

This very well looked after Mk.10 Chieftain belongs to the Belgian Tank Museum, and is a credit to them. It regularly has days running for the general public, and is a joy to see and hear [Tank Museum, Brussels].

Another view of the recovery in place, scenes like this are valuable for a modeller as they often show areas that are not normally accessible to the general public [Plain Military].

This rather dusty Chieftain is located at the Imperial War Museum at Duxford, the view shows the top layout very well of interest the green boxes on the NBC pack are the units that are required to make the SIM-FICS system work. Also the detectors can be seen around the turret with the flash/bang generator to the right of the gun. Unlike SIMFIRE the laser is located inside the barrel at the muzzle end rather than on top of the gun [R. Griffin].

Rather beyond restoration possibly this hulk is slowly being battered to scrap on the ranges, Chieftain has proved a good hard targe[t]
and many have extra armour plates lent against them to prolong their life [R. Griffin].

Walkaround – R. Griffin,
Courtesy of RMCS Shrivenham AFV Wing

CHIEFTAIN
MBT

In 1983 the Chieftain Mk.11 conversion was approved to fit TOGS to a portion of the Chieftain fleet, and 324 vehicles were converte[d]
at base workshops from existing vehicles held in the UK and at BAOR. The Mk.11 and the less advanced Mk.10 supplemented the Chal[-]
lenger 1 Main Battle Tank in service in the Federal Republic of Germany (then known as West Germany) until 1994, but many soo[n]
found their way to BATUS and to the UK bases as more Challengers entered service. Some of the Mk.11 conversions started life a[s]
Chieftain Mk.2s that had started their service in the 1967-69 period, while others were converted from Mk.3 series and Mk.5 vehicle[s.]
The Chieftain's chequered mechanical history meant that most of the fleet had already undergone substantial modification by the earl[y]
1980s as part of the Sundance program and the Mk.11 conversions were based on previously upgraded tanks. These "late" Chieftain[s]
(which were all rebuilt "early" Chieftains) were far less prone to mechanical problems than they had been as configured at the star[t]
of their service, and TOGS and the new APFSDS munitions developed for the L11A5 gave the Chieftain a new lease on life. The las[t]
Chieftain regiment in the Royal Armoured Corps were the 1st RTR in 1994, by which time the Chieftain's steel armour was obsolet[e]

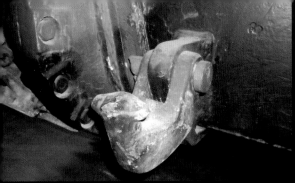

The front towing points featured fixed shackles for vehicle recovery.

A close-up of the front driving lamp cluster on the front left of the glacis plate. A mirror image of this arrangement was used for the opposite side, with the infra-red lamp in the inside and the regular white headlamp in the outside position, under a wire brush guard.

Later Marks of Chieftains carried rubber mud flaps bolted onto the front track guard.

The Chieftain's drive sprocket.

The front idler mounting seen from underneath the tank.

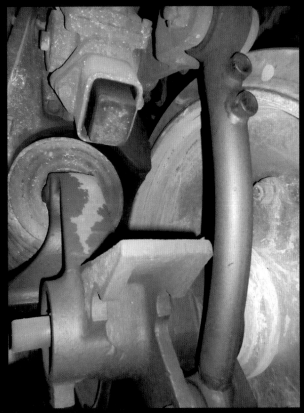

Detail view of a Horstman suspension unit.

The rear of the Chieftain carried a running light on each side, here is a close up of the type employed.

Shock absorber bump stops are fitted to the front Horstman suspension unit on each side.

The gun clamp, introduced with the L11A5 120mm rifled gun fitted to the Chieftain Mk.5 in 1972. This gun clamp was retrofitted to nearly the entire British Army Chieftain fleet, and certainly featured in most vehicle rebuilds and upgrades (as did the new gun

The Convoy Marking seen here (and lit by the small light above it) was not employed by all units, some of whom employed a simple white rectangle with the unit code or vehicle call sign painted inside in black.

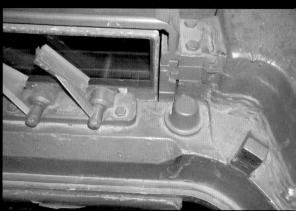

The fire extinguisher pull handle located between the front stowage boxes on each side of the Chieftain's track guards.

The driver's periscope with its wiper blades was located just behind the driver's hatch in the middle of the glacis plate.

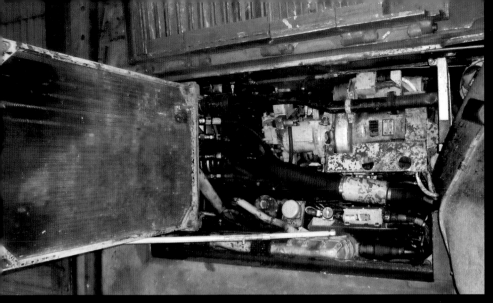

Looking down onto the H30 Auxiliary-Generator, which powered the tank's electrical systems when the main engine was not running.

Looking down on the L60 main engine air cleaner, the three round objects are the main engine oil filters.

The vertical object is one of two radiators, while the open round cover is the access point for the engine coolant level.

In this view can be seen the auxiliary generator oil filler (silver cap) and the round object is the air cleaner, with an indicator on the top showing when it needs cleaning.

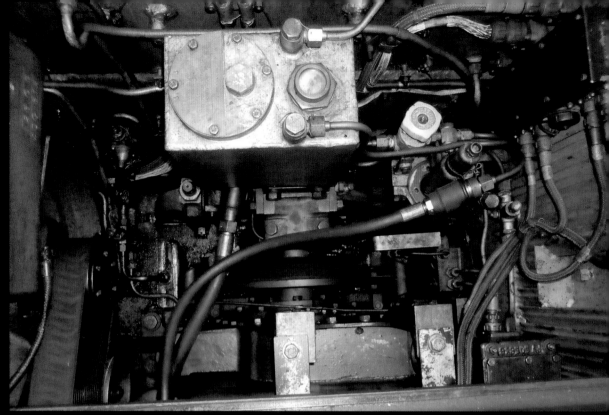

This is the hydraulic brake fluid reservoir and the view shows just how cluttered a modern MBT hull is.

This shot is the lower part of the large stowage basket on the turret left side.

This the air intake for the NBC pack on the turret rear, the handle half way up and the two knobs at the bottom control two blanking flaps that must be closed prior to washing the vehicle.

A close up of the right hand side of the turret showing the mass of Stillbrew armour protecting the turret face. The conduit visible is for the electrically fired smoke discharger unit, and we can clearly see the Clansman wireless box marked with the insignia of the AFV Wing, and behind it the commanders stowage box first introduced on the Chieftain Mk.3.

A six round smoke discharger is fitted on a bracket on each side of the turret's frontal armour.

The basic design for these smoke dischargers seems to have also been adopted by the US Army for the M60 and M1 series of MBTs, but in the British Army has long been superseded by a newer type introduced on the Challenger 1.

The gunners' main sight with its armoured cover in the raised position.

The gunners' secondary sight was a simple fixed periscope in the event of combat damage to the main sight.

The loader's periscope was traversable and offered a useful third set of eyes to the turret crew on observation.

The TOGS (Thermally Operated Gunnery System) barbette was a lightly armoured structure added to the left side of the Chieftain's turret, replacing the infra-red searchlight and much of the turret outside stowage. Developed for the Challenger 1, TOGS was successfully retrofitted to the Chieftain, but its layout was not ideal in terms of turret ergonomics. Nonetheless the Mk.11 was the most capable Chieftain gun tank and was quite at home night fighting. The door here is in the closed position.

A top view of the bracket which held one of the two turret rear stowage boxes added to the Mk.11 (and going by photos, to most Mk.10s as the new boxes became available) to compensate for the stowage space lost to the TOGS system.

This one of a series of hooks that were used to hold the support wires that held the wading tower, which was developed for snorkelling operations for deep fording: although trials were successfully completed the tower was never issued to units.

This lightly armoured box contains the TUAAM (Tuning Unit Automatic Antenna Matching): this unit allowed the 2m whip antenna to be tuned electronically to the correct frequency.

One variant of the Clansman radio antennae base, normally a 2m whip would be fitted but sometimes to reduce the silhouette it would be replaced by 1m, with a corresponding decrease in communicating range.

This shows the rubber bellows and the canvas mantlet cover fitted to ensure that no part of the barrel was exposed to the elements. Replacing the mantlet cover was one of the most hated maintenance chores.

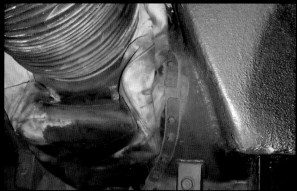

This is an access hatch for maintenance on the TOGS system.

In this view the original cast turret can be seen along with the depth of the Stillbrew add on armour. The Stillbrew armour provided added protection against the Soviet 125mm armour piercing round and against infantry antitank weapons. It was introduced in the 1982-83 period.

This shows the rear portion of the smaller fume extractor fitted to L11A5 120mm barrels introduced with the Chieftain Mk.5 and retrofitted to earlier marks in base overhauls. The two elements at the rear are counterweights.

The forward edge of the L11A5 fume extractor: clearly seen are the very large clamps that secure the thermal sleeve to the fume extractor as well as the rearmost thermal sleeve clamp.

This is one of 9 wiper motors located around the cupola, notice how it is angled so that it matches the slope of the periscope face, the wipers are controlled from inside the cupola by the commander.

Located on the left of the GPMG (General Purpose Machine Gun) mount is the original style of ammunition box holder, which would only hold one box of 7.62mm ammunition. Later variations had ammunition stowage fixed to the cupola, enabling more rounds to be stored and one of the last variants employed on the No. 15 Cupola was a large curved ammunition bin located on the left side of the cupola top.

The commander's spotlight could be fitted with an infra-red filter if required, although its main purpose on exercise was to be flashed on and off to indicate that the tank had "fired" its 120mm as an indication to the "enemy".

This locally produced modification was used to hold the status flags, used when on the gunnery range. Normally a red flag was shown for firing and green was shown for all weapons clear, other issues required a combination of flags, which explains the two holders!

The chute on the right of the GPMG was to direct the spent links from the ammunition belt away from the machine gun mounting. The belt of ammunition for the GPMG is of the disintegrating link type and unlike a cloth belt the empty cases fall from the bottom of the gun while the links fall from the right side. Keeping the spent brass and links from cluttering the turret and potentially causing problems was an important consideration.

The turret roof from the loaders' side, showing the basic layout of the hatches and the loader's periscope. The facility of contra-rotating the cupola and turret was dispensed with on the Mk.11 due to the space occupied by the TOGS system.

The loader's hatches inside faces are padded and incorporate rubber seals for operation in an NBC environment.

The levers shown operate to open or securely lock the hatch.

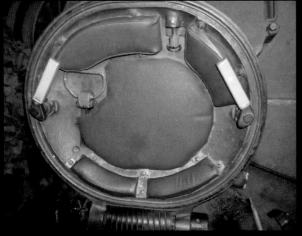

The commanders' hatch opens towards the rear of the cupola with the cupola traversed to its normal position, and can be locked into 1 of 3 positions depending on the tactical situation.

A view down through the drivers hatch reveals the most comfortable seat of any crew position on the Chieftain. The seat was adjustable so that when closed down the driver was in a supine position, almost lying down, but when the driver's hatch was open he sat up in a regular seated position.

A good shot of an operational drivers cab, with foot pedals in the centre and on the right the aux gen throttle and emergency gear levers, on the left the lever for the hydraulic start system, and in the center can be seen the instrument panel.

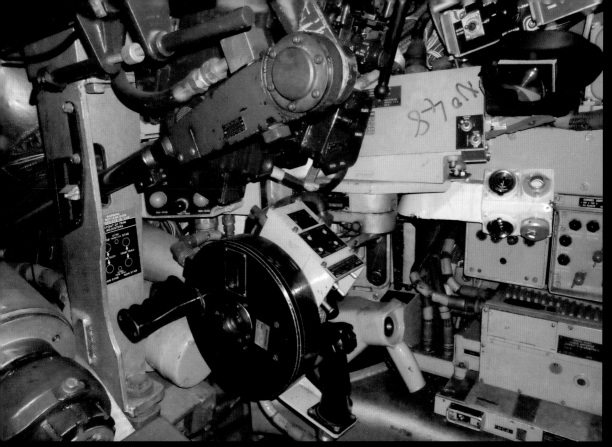

Looking from the loaders side across to the gunners station, at the top right of the picture is the gunners TOGS viewer, with the large round wheel for hand elevation central, and to the right is the gunner's fixed firing grip.

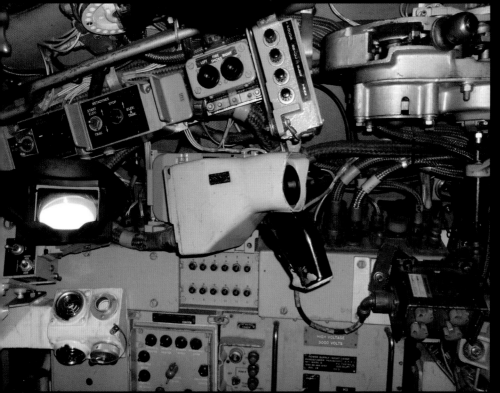

Central in this shot is the commanders thumb controller, which was used to control turret power traverse and elevation from the commander's station; it could override the gunners' controls which were identical. To operate, the lever at the front was pressed in and the commander used his thumb to operate the small black spring loaded control on the forward edge of the unit in the picture.

To the left of this shots is original contra –rotation controls, the large gray box to the right is the Commander's control and monitoring unit, this gives him data on range, faults in the system and allows him to make corrections when firing.

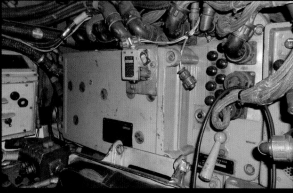

Behind the large panel pictured are located the various removable electronic control units that provided the power system for turret traverse and elevation. If a fault occurred it was usually easier to remove faulty unit and plug in a new one than attempt a local repair.

This component of the TOGS system had one of the more colourful names used, correctly it is the symbology processing unit but more commonly called spu!! It control many function of the TOGS system, and when not in use it was protected by the lid visible in the picture.

This is the turret services electrical distribution box, various electrical functions were provided by this unit, including the facility to link all the vehicle batteries together. This would allow the tank to be fought under battery power if all power generating facilities has been lost, if only for a very limited period. Perhaps its most important service was providing power to the boiling vessel.

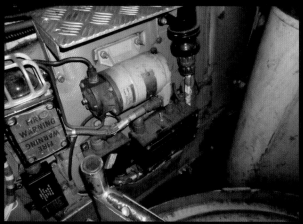

Located on the hull wall is one of the vehicle washer pumps and next to it can be seen part of the fire alarm system.

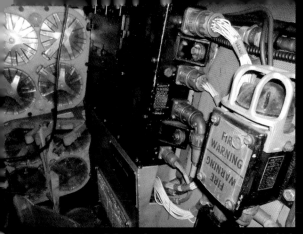

The fire alarm flashing light, in the event of a fire the alarm (which was incredibly loud) rang, and the red light flashed. The system was known to have faults and cause false alarms, which were always interesting occasions!

All British military vehicle have a variation of this plate, known as a census plate. On it can be found all the details relevant to that vehicle, including date of build, workshop repairs and mark.

The commander's station, showing in the centre his X10 Sight No. 37 Mk.3. The black handle to the centre left is the control for firing his cupola GPMG, and below that is his fixed firing grip which allows him to fire main armament, the co-axial machine gun, the tank laser sight and to use the full IFCS (Improved Fire Control System).

Painted by Sławomir Zajączkowski

The 15/19th Kings Royal Hussars received their first two Chieftain Mk.2 in 1969 at Tidworth, one of which was 04 EB 80 (call sign 33D). The regiment had previously been equipped with Centurion Mk.5/2, Mk.8 and Mk.10. Photographs show these early production Chieftain Mk.2s with the large hull rear stowage bins, without driver's splash boards on the glacis or stowage baskets on the right side of the turret bustle. The tanks were almost certainly painted Deep Bronze Green and the 120mm gun barrels were painted white from the rear edge of the fume extractor forward to the muzzle. The marking system is quite colourful, and seems to have been unique to this regiment. The RAC flash superimposed with the unit serial 37 in white paint has not been seen elsewhere. The marking was

unusual in that it used a spare code number allocated temporarily to the brigade that the 15/19th KRH were assigned at Tidworth for a year prior to deployment to BAOR, when they still were equipped with Centurions (and probably worn from 1968 to 1970). The regimental crest was applied on the Infra Red searchlight housing door centered towards the lower edge. The call sign was painted on the second Bazooka plate and possibly on a plate fixed to the turret rear. The 15/19th KRH took over their issue Chieftains from the Royal Scots Greys when they arrived in Germany. (with thanks to Major Mel Tazey)

This is an early Chieftain Mk.2 04 EB 09 of the Blues and Royals when they converted to the type in 1969. The markings are recorded from several well known photos from 1968-1969 when the Chieftain was the pride of the Royal Armoured Corps. The tank depicted call sign 0B had early features and lacked the glacis splash board that became almost universal on Mk.2s as the 1970s progressed. Markings are quite colourful in comparison

to later Chieftains: the white painted smoke dischargers being a unique feature to a few particular tanks, for reasons unknown and was not a Blues and Royals speciality. The camouflage is BAOR Black and NATO Green disruptive pattern applied with smart sharp edges and the unit code of 20/2(which had replaced the old arm of service markings in the 1960s) is visible. The yellow bridging symbol and the regimental crest on the Infra-Red Searchlight housing were typical for the time but disappeared by the late 1970s. In 1971 the Blues and Royals handed over their Chieftains to the Life Guards and left West Germany to re-equip with CVRT as a reconnaissance regiment.

Painted by Sławomir Zajączkowski

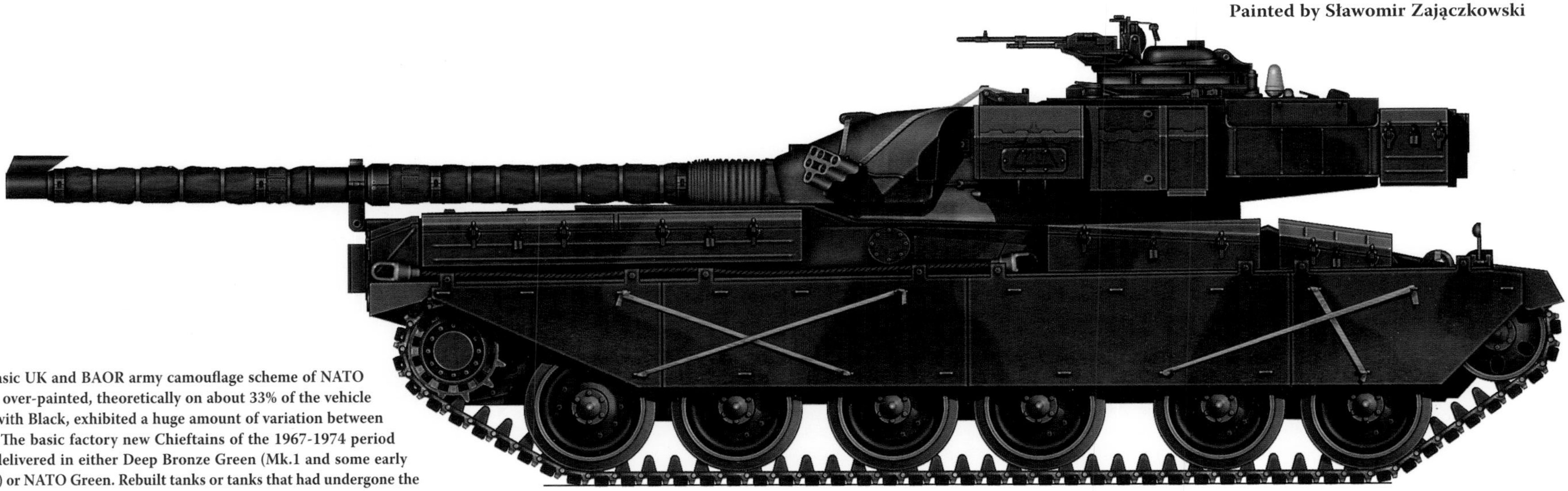

The basic UK and BAOR army camouflage scheme of NATO Green over-painted, theoretically on about 33% of the vehicle area, with Black, exhibited a huge amount of variation between units. The basic factory new Chieftains of the 1967-1974 period were delivered in either Deep Bronze Green (Mk.1 and some early Mk.2s) or NATO Green. Rebuilt tanks or tanks that had undergone the long series of upgrades were usually repainted as part of the process, and were delivered back to their units in NATO Green. At the unit level the patterns for the black disruptive colour were completely at the discretion of the troop painting them, with each troop having its own ideas on patterns. At the same time the Regimental markings exhibited a minimalist trend throughout the later 1970s and 1980s in BAOR. The two anonymous Chieftains pictured here are typical of the Black and NATO Green paint scheme in the mid 1980s and show exercise markings as commonly applied during the large fall exercises conducted with allied NATO forces. Note also that during exercises the bazooka plates were sometimes removed to alter the Chieftain's profile and make it more resemble an *enemy* tank.

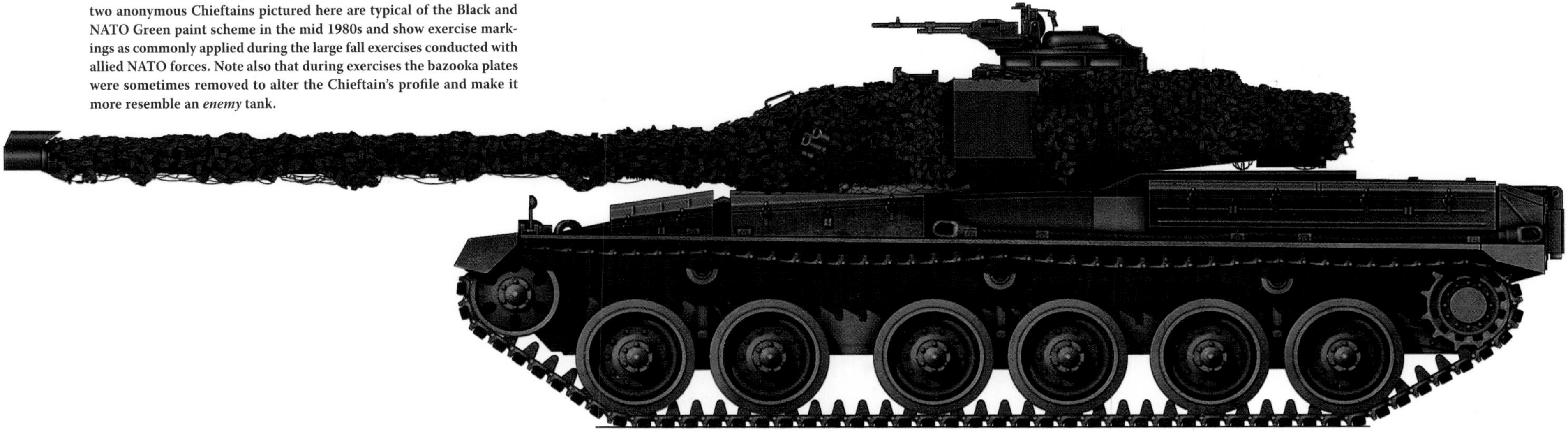

Painted by Sławomir Zajączkowski

The 4th RTR was stationed at Tidworth, U.K. in 1984, and the colour plate shown depicts *Dewar* (05 EB 49), one of the Mk.5, Mk.6 and Mk.7 Chieftains the regiment had on strength at the time. Other names in use at the time were *Dolly* (call sign 32), *Denali* (possibly 01 EB 45), and the command tank *Royal Sovereign* (00 FD 55). Most of these tanks had the Muzzle Reference System fitted and large red squares painted on the bazooka plates for exercise purposes. The standard markings used by UK Chieftain regiments by this time were very spartan compared to those of the 1970s. Bridging markings were changed to

grey backgrounds, convoy markings were simplified to a white rectangle on the rear plate, and call signs usually only appeared on the turret rear. Only the Union Jack on the front mudguard gave much colour. Most famous of the 4th RTR's markings were the Chinese eyes on the turret front, a practice that dated back to the First World War. (based on photos taken by Tim Neate)

The Mk.11 depicted below is from either the 4th RTR or the 1st RTR following its amalgamation with the 4th RTR in 1989 as the Royal Armoured Corps began to shrink with the end of the Cold War. The Chinese eyes were retained as a 1st RTR tradition and continue to be so. The 1st RTR was the last regiment to operate the Chieftain gun tank, its last Mk.11s being turned in to ordnance in 1994, by which time the Chieftain's steel armour was obsolescent. The TOGS system introduced with the Mk.11 was not an ideal ergonomic fit in the Mk.11 and required the removal of the cupola contra-rotation equipment, but gave the Chieftain a solid night fighting capability. By the late 1980s the use of fluorescent hazard markings on the rear of BAOR AFVs was widespread due to West German road laws, and on the Chieftain these consisted of two panels attached to the rear hull stowage bins.

Painted by Sławomir Zajączkowski

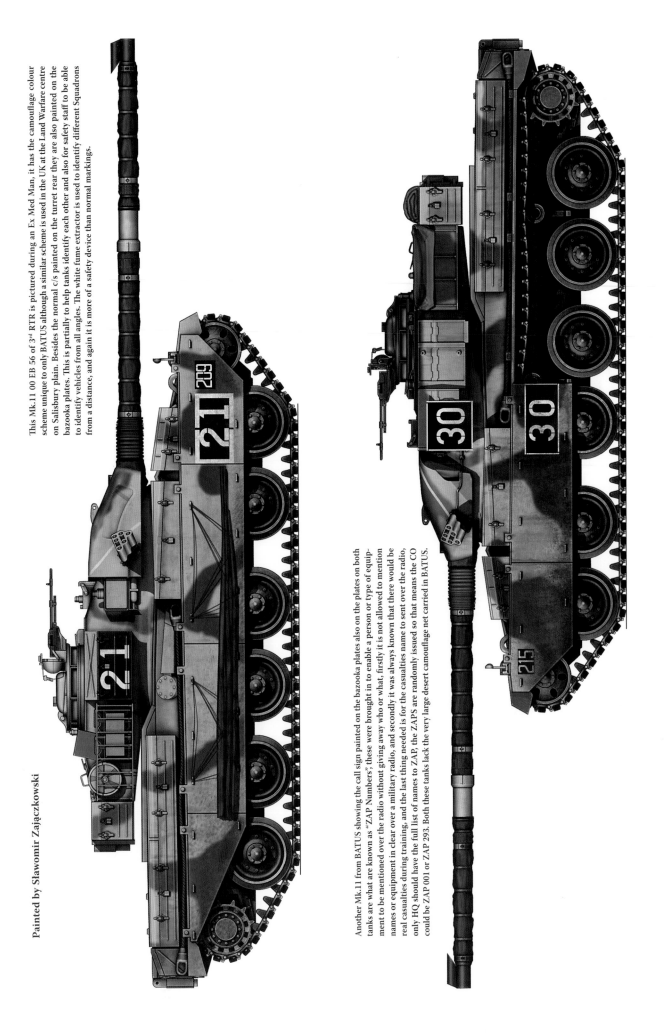

This Mk.11 00 EB 56 of 3rd RTR is pictured during an Ex Med Man, it has the camouflage colour scheme unique to only BATUS although a similar scheme is used in the UK at the Land Warfare centre on Salisbury plain. Besides the normal c/s painted on the turret rear they are also painted on the bazooka plates. This is partially to help tanks identify each other and also for safety staff to be able to identify vehicles from all angles. The white fume extractor is used to identify different Squadrons from a distance, and again it is more of a safety device than normal markings.

Another Mk.11 from BATUS showing the call sign painted on the bazooka plates also on the plates on both tanks are what are known as "ZAP Numbers", these were brought in to enable a person or type of equipment to be mentioned over the radio without giving away who or what, firstly it is not allowed to mention names or equipment in clear over a military radio, and secondly it was always known that there would be real casualties during training, and the last thing needed is for the casualties name to sent over the radio, only HQ should have the full list of names to ZAP, the ZAPS are randomly issued so that means the CO could be ZAP 001 or ZAP 293. Both these tanks lack the very large desert camouflage net carried in BATUS.

Painted by Sławomir Zajączkowski

The 3rd RTR showed a great deal of paint scheme variation within the Regiment itself. The tanks of A Squadron at Sennelager, West Germany, in 1982 were painted with a scheme which gave the impression of the black pooling along the lower edges of the bazooka plates, as seen on 06 FA 72 depicted here (call sign 12B). At the same time C Squadron used a soft edged camouflage of wavy lines and the regiment's headquarters tanks used a completely unique camouflage scheme of black blotches. The unit codes seem to have been kept in the 3rd RTR until the mid 1980s, though other markings were minimal other than the squadron symbols, union jack, VRN and bridging markings. By the time the 3rd RTR was about to convert to Challenger 1 MBTs in the mid 1980s, the Chieftain's turret markings only consisted of the call sign in white on the rear of the turret.

BATUS, or British Army Training Unit Suffield, was established at CFB Suffield (in Alberta, Canada) to enable the Royal Armoured Corps to establish a battle training center away from the restrictions of the West German and British training areas. The wide open spaces of Suffield enabled large operational areas and umpiring the many exercises soon gave rise to distinctive markings visible at long range. Each tank was eventually given a prominently painted zap code in addition to its call sign, and the turret received white stripes at 22.5 degrees each side of the main gun to act as a safe area indicator (the armaments not being allowed to fire if anyone could be seen within these lines, which were known as the 45s). BATUS's tanks stayed at the base, but individual vehicles had to be periodically be rotated back to the UK for rebuilds. This Mk.11 was probably there in the late 1980s as the availability of Challenger 1 allowed some modernized Chieftains to be spared from BAOR and UK armoured regiments.

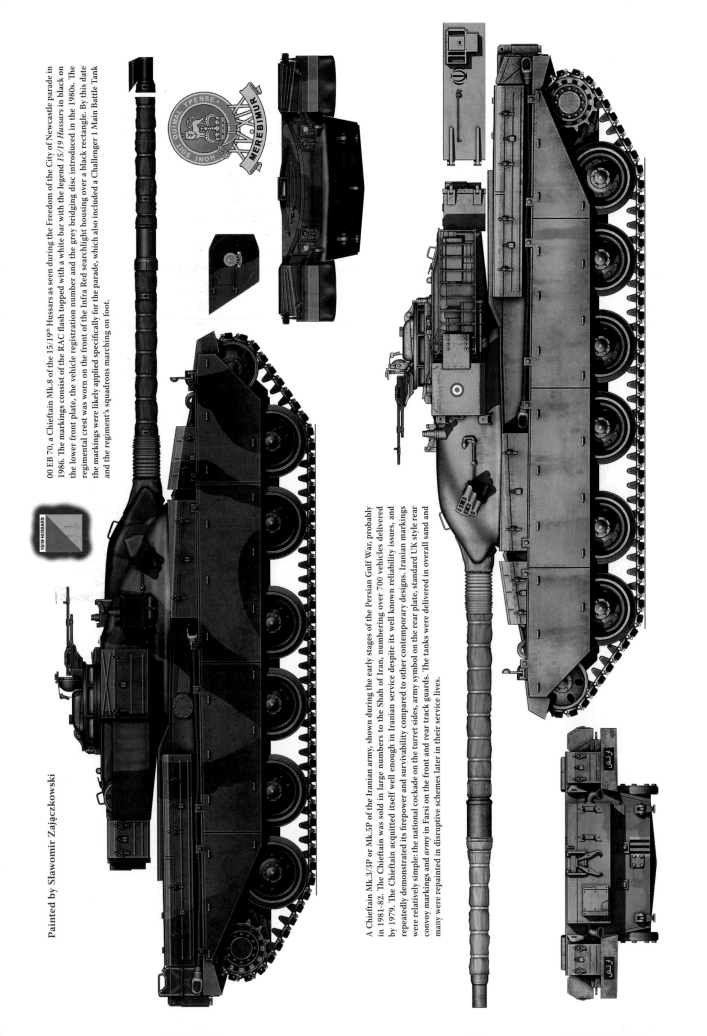

Painted by Sławomir Zajączkowski

00 EB 70, a Chieftain Mk.8 of the 15/19th Hussars as seen during the Freedom of the City of Newcastle parade in 1986. The markings consist of the RAC flash topped with a white bar with the legend *15/19 Hussars* in black on the lower front plate, the vehicle registration number and the grey bridging disc introduced in the 1980s. The regimental crest was worn on the front of the Infra Red searchlight housing over a black rectangle. By this date the markings were likely applied specifically for the parade, which also included a Challenger 1 Main Battle Tank and the regiment's squadrons marching on foot.

MEREBIMUR

A Chieftain Mk.3/3P or Mk.5P of the Iranian army, shown during the early stages of the Persian Gulf War, probably in 1981-82. The Chieftain was sold in large numbers to the Shah of Iran, numbering over 700 vehicles delivered by 1979. The Chieftain acquitted itself well enough in Iranian service despite its well known reliability issues, and repeatedly demonstrated its firepower and survivability compared to other contemporary designs. Iranian markings were relatively simple: the national cockade on the turret sides, army symbol on the rear plate, standard UK style rear convoy markings and *army* in Farsi on the front and rear track guards. The tanks were delivered in overall sand and many were repainted in disruptive schemes later in their service lives.

Painted by Sławomir Zajączkowski

A typical Chieftain Mk.5 dozer, shown in early BATUS markings in the 1970s (prior to the introduction of zap codes into the marking scheme). The dozer was usually fitted to the squadron second in command's (or 2IC's) tank and was a standard issue conversion kit. The dozer blade was useful to scrape out fire positions or to improve river fords but was never a popular piece of equipment due to hydraulic failures common to the dozer equipment. One of the most common uses for the dozer in BAOR, especially during the mid 1970s when vehicle running hours on training were restricted as a cost cutting measure in BAOR, was as a base snow plow!

Cambrai, was the Commanding Officer's tank of the 3rd Royal Tank Regiment, seen here in 1979 at Hohne training area wearing a blotch pattern camouflage, an extreme and striking variation of the BAOR Black and NATO Green scheme. The unusual paint scheme by all accounts delighted the 3rd RTR commanding officer, a true testimony to the fact that a *standard* BAOR black and green scheme was completely mythical. The regimental code 11/3 may have been painted on the tank's front plate next to the bridging disc. The striped convoy marking was still retained on this vehicle's rear plate at the time it is shown, the 3 black stripes superimposed on the convoy marking in the 1970s having been simplified to a plain white rectangle by 1980 in many regiments, including 3rd RTR). Other markings included the HQ Squadron diamond painted in white on the commander's stowage box and on the outside of the Infra-Red searchlight housing. The vehicle call sign ØD was painted in white on a black plate attached to the turret rear. The Union Jack decal may have been applied to the rear left mudguard on *Cambrai*, which have carried it on the front left track guard as well. The vehicle registration number 11FD33 (denoting a Mk.5) was applied on the right rear mudguard. This tank carried jerry can racks made up by the regimental fitters on the sides of each rear stowage bin. (with thanks to Ken Holland and Bob Jacobs)

Painted by Sławomir Zajączkowski

This Chieftain is representative of a number of Iranian Chieftain Mk.5P tanks found in Iraq in 2003, and was probably part of the equipment of a unit of tanks crewed by Iranian counter-revolutionary troops fighting in Sadam's Iraqi army. These tanks were captured and reconditioned, most likely due to defections by the crews or as a result of capture by Iraqi forces due to mechanical failure. The poor level of training many Iranian armoured corps conscripts received following the 1979 revolution had serious consequences on a maintenance-intensive weapon system like the Chieftain. These tanks were painted in basic sand with green over sprayed stripes and carried variations on the usual Iranian markings. It is probable that the tanks had been sitting since the end of the Iran-Iraq war and it is unknown if they saw extensive combat against their former owners.

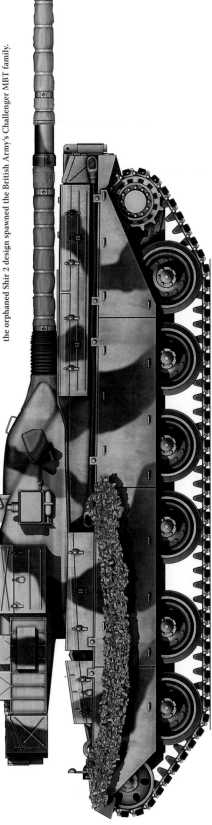

This Iranian Chieftain Mk.5P shows one of the many colour schemes worn during the Iran-Iraq war. The Chieftain was actively marketed to Belgium, the Netherlands, India and Lybia (to name a few), but other than the large orders placed by Iran, only Kuwait and Oman bought the type. The Iranian orders were to be followed by the much improved design of the Shir 1 and Shir 2 MBTs, but once the Islamic revolutionary government of Ayatollah Khomeini cancelled all arms contracts with Britain the economic disaster that would have resulted for the British tank industry required decisive government action. The Shir 1 became the Jordanian Khalid MBT and the orphaned Shir 2 design spawned the British Army's Challenger MBT family.

Painted by Sławomir Zajączkowski

The Berlin Armoured Squadron was a reinforced squadron of tanks crewed on a rotating basis by squadrons detached from the armoured regiments of BAOR and had roots extending back to the Berlin Crisis of 1948. The British component of the West Berlin Garrison was known as The Berlin Brigade. The exotic urban scheme worn by the AFVs of the Berlin Brigade was developed in the early 1980s when the 4/7th Royal Dragoon Guards D squadron was detached to the garrison. The scheme consisted of irregular squares and rectangles of Olive Drab, Grey and White, with wheels painted black. By the time the scheme was introduced the Mk.5, Mk.6, and Mk.7 Chieftains were in service with the Berlin Armoured Squadron.

The 14/20th Hussars were the last regiment to man the Chieftains of the Berlin Squadron, departing in 1989 (and going on to fight in Challenger 1s in *Operation Granby* shortly thereafter). This Mk.10, 02 EB 20, began life as a Mk.2 and was displayed at Imperial War Museum Duxford in 1999. It is shown here as it would have looked in the late 1980s. The vehicle hull rear stowage bins carried the standard West German fluorescent traffic panels and the 14/20th Hussars Regimental Crest was applied on an yellow (or light orange) rectangle on the side and front of the Infra Red searchlight housing. The Mk.9, Mk.10 and Mk.11 conversions also often included the extensive provision of mesh screens over the engine deck grilles, and were fitted to the Mk.10 pictured.